ROUTLEDGE LIBRARY EDITIONS:
ENVIRONMENTAL POLICY

Volume 5

T0175170

RISKS AND OPPORTUNITIES

RISKS AND OPPORTUNITIES
Managing Environmental Conflict and Change

VALERIE BROWN, DAVID INGLE SMITH,
ROB WISEMAN AND JOHN HANDMER

Routledge
Taylor & Francis Group

LONDON AND NEW YORK

First published in 1995 by Earthscan Publications Ltd

This edition first published in 2019
by Routledge
2 Park Square, Milton Park, Abingdon, Oxon OX14 4RN

and by Routledge
52 Vanderbilt Avenue, New York, NY 10017

Routledge is an imprint of the Taylor & Francis Group, an informa business

British Library Cataloguing in Publication Data
A catalogue record for this book is available from the British Library

ISBN: 978-0-367-18894-8 (Set)
ISBN: 978-0-429-27423-7 (Set) (ebk)
ISBN: 978-0-367-22173-7 (Volume 5) (hbk)
ISBN: 978-0-367-22185-0 (Volume 5) (pbk)
ISBN: 978-0-429-27366-7 (Volume 5) (ebk)

Publisher's Note
The publisher has gone to great lengths to ensure the quality of this reprint but
points out that some imperfections in the original copies may be apparent.

Disclaimer
The publisher has made every effort to trace copyright holders and would welcome
correspondence from those they have been unable to trace.

RISKs

AND

Opportunities

MANAGING ENVIRONMENTAL CONFLICT
AND CHANGE

Valerie Brown   David Ingle Smith
Rob Wiseman   John Handmer

EARTHSCAN

Earthscan Publications Ltd, London

First published in Great Britain in 1995 by
Earthscan Publications Ltd, 120 Pentonville Road, London N1 9JN

A catalogue record for this book is available from the British Library.

ISBN: 1 85383 236 7

Earthscan Publications Limited is an editorially independent subsidiary of Kogan Page Limited and publishes in association with the International Institute of Environment and Development and the World Wide Fund for Nature.

Printed in England by Biddles Ltd, Guildford and Kings Lynn
Cover by Yvonne Booth

Contents

6

List of Figures and Tables

Figures

Tables

Glossary

BMA	Bangkok Metropolitan Administration
BMR	Bangkok Metropolitan Region
CO	carbon monoxide
CRES	Centre for Resource and Environmental Studies
DFO	Department of Fisheries and Oceans
DIW	Department of Industrial Works
EIA	Environmental Impact Assessment
FREMP	Fraser River Estuary Management Program
IAC	Industries Assistance Commission
IDO	Interim Development Order
IIFC	Interior Indian Fisheries Commission
IUCN	International Union of Conservation Organisations
MDBC	Murray-Darling Basin Commission
NBCC	Native Brotherhood of British Columbia
NESDB	National Economic and Social Development Board
NGO	Non-Government Organisation
NIMBY	'Not In My Back Yard'
OECD	Organisation for Economic Cooperation and Development
OTCP	Office of Town and Country Planning
PFDA	Pacific Fishermen's Defense Alliance
RWC	Rural Water Commission
SEC	State Environment Council
SRWSC	State Rivers and Water Supply Commission
UFAWU	United Fisherman and Allied Workers Union
UNCSD	United Nations Commission for Sustainable Development
UNDP	United Nations Development Program
WAC	Water Advisory Committee
WCED	World Commission on Environment and Development
WHO	World Health Organisation

Acknowledgements

In 1990, the Centre for Resource and Environmental Studies (CRES) at the Australian National University, Canberra, was funded by the National Universities Priority Reserve Fund to develop a course in environmental conflict management, using case studies in Australian environmental management. The teaching of this course, and the encouragement of the participants, provided the original stimulus for this book. Professor Henry Nix, director of CRES, gave us sound advice and support during the three years that it took to further develop the teaching materials and case studies.

The central themes were:

- managing environmental change will inevitably mean managing conflict – between community interests, professional positions and political priorities, and the present and the future, as well as between conservation and development;
- the outcome of the conflicts can be constructive or destructive, depending on the management; and
- successful conflict management minimises the risks and maximises the opportunities for both conservation and development.

The learning programme was tested twice, with participants who were directly involved in Australian environmental management: managers, administrators, academics, conservation groups, negotiators and students. They were working in water management, public health, education, law, engineering, politics and mediation. Their experience and advice contributed a great deal to the development of the materials. Their names are listed in Appendix 2; and the authors gratefully acknowledge their contribution.

The case studies used to illustrate the *Why*, *What*, *How* and *Who* of integrated environmental management have been developed from existing work, with the assistance of the primary authors. These are acknowledged in the relevant chapters. We would particularly like to thank Tony Dorcey of British Columbia,

11

Anuchat Poungsomlee of Bangkok, Bruce McKenzie of the Murray-Darling Basin and Ed Miller of Maryborough for both their technical assistance and for sharing their deep understanding of their own people and place.

The exercises in Part 3 incorporate the fruits of many years of experience in professional education. Some of the teaching materials have been adapted from those in the public domain (for instance, those of Hugh Watson, Canberra, from his Structured Experiences) and some have been designed especially for this manual. For many of these, we are grateful for the considerable expertise of consultant Ian McAuley, of 24 Schlich Street, Yarralumla, Canberra, who, with the assistance of Linden Orr, Nora Stewart and Anna Carr, helped us teach the second of the two courses.

McComas Taylor, in conjunction with Rob Wiseman, edited the teaching materials and designed the format of the original teaching manual with his considerable flair. He also provided an essential confidence in the manuscript and personal moral support for the authors. It is only because McComas and Ettie Oakman man-and-woman-handled the text into readable form not only once, but twice, (once for the teaching manual, and once for the publisher) that the text finally saw the light of day. Thank you both.

Most of all, we would like to acknowledge the key role of Linden Orr, who played a major part in designing and presenting the teaching programme, and in the preparation of two of the case studies. She also practised what we were preaching, using her own skills in conflict management to keep the team moving forward. Although she had moved on by the time we came to put the book of the course together, the form and shape of the materials still bear her imprint.

We accept full responsibility for the final form of this book, and would be delighted to hear from anyone applying these materials, in order to refine them further.

Valerie A Brown, D Ingle Smith,
Rob Wiseman, John Handmer

Centre for Resource and Environment Studies
Australian National University
Canberra, May 1994

Part 1

Managing Environmental Conflicts

1   **Opportunity and Risk**

Half Full or Half Empty?

The human species may be half way to ruining our planet – or we may be half way to harnessing its rich resources. In the industrialised countries, at least, we tend to regard the two statements as mutually exclusive. But they are not irreconcilable. In fact, they are both true at the same time. Yet whether we are moving 'forwards' or 'backwards', progressing or regressing, developing or destroying, is a matter of endless and often angry debate. The protagonists might just as well argue over whether a glass of water is half full or half empty. The conflicting positions hinge on different expectations of exactly the same conditions:

- On the one hand, we cannot afford to take risks with the ways in which we manage our environment, since it is the human species' only habitat. Without it we have no future.
- On the other hand, we must make the most of our opportunities to maximise the world's resources, for our own good and for those who come after us.

Both responses indicate concern for the continuation of the earth's living systems and of its human burden. Both positions are argued with equal conviction on behalf of this generation (ourselves and our children) and the next (our children's children). But the argument between the two positions – positions not in themselves inherently opposed – is at risk of taking priority over constructive action on either one.

The environmental debate may be said to have begun in earnest when the World Commission on Environment and Development sent up their smoke signals that everyone, not only those who are ecologists and conservationists, bears responsibility for the condition of the natural resources of the globe (WCED 1987). The Commission drew the world's attention to the fact that the

14

rate of human resource use now exceeded the earth's self-managing capacity, and its management had now become a matter of human intervention. The International Union of Conservation Organisations came to the same conclusion four years later (IUCN 1991). Since then, many interests have accepted some responsibility for monitoring the condition of the planet; only each of them is monitoring a different aspect, often from a different point of view. Decisions based on both the risks and the opportunities for environmental management become more difficult, when even the very sources of information are polarised.

To the regular reviews of the health of the global economy by the OECD and the World Bank have been added the annual reports on the state of the environment from non-government organisations such as the Worldwatch Institute (Brown 1989–93). Since the 1992 World Conference on Environment and Development, the United Nations Commission for Sustainable Development has taken the responsibility for reports on progress towards Agenda 21, a blueprint for global environmental management for the 21st century (CSD 1994). Countries report annually on their progress towards Agenda 21, and on the world conventions on biodiversity and climate change. Local authorities world wide conduct their own monitoring of economic development and environmental management (usually separately) for their own purposes, at the local scale. Streams of reporting on the states of the world's economy and environment are now in place at the global and the local level. Managing the economic and natural aspects of the one global environment has been set up in parallel, and often opposing, streams of decision making.

When government and non-government organisations monitor the same environments, the way in which the balance is calculated can be very different. The Club of Rome report *Limits to Growth*, which blew one of the first danger whistles, has been brought up-to-date with more optimism than in 1972, noting some wins and some losses (Meadows et al 1992). Control of pollution is up; but so is the worldwide scale of pollution. Nuclear war did not eventuate and new nuclear power facilities are slowing down; but the nuclear wastes from military and power generation are accumulating with no disposal solution in sight. Overall, the review of all the evidence leads Meadows et al to conclude:

- the human rate of use of many essential resources and generation of many kinds of pollutants is no longer physically sustainable;
- the decline is not inevitable, if there were a comprehensive revision of growth policies, and a rapid increase in efficiency of use of materials; and
- a sustainable society is still technically and economically possible.

15

The Worldwatch *State of the World* results for 1993 make for depressing reading. They continue the record of the spread of cities over productive land; degradation of global water cycles, soil and forest resources; and the exponential population growth. More importantly, the same authors have offered some hope of an effective, long-term response (Brown 1991). There has been some meeting of minds between the polarised 'half full, half empty' positions (implying 'keep going, there's plenty left' as opposed to 'stop everything, we are running out') on biodiversity, climate change and the state of the ozone layer. The predictions of climate change and of the need to prevent global warming and ozone depletion are formally accepted by most countries (UNCED 1992). As these conventions were being prepared, Mt Pinatubo in the Philippines sent out a cloud of aerosols, greatly adding to the risk of ozone loss. Just as those from the 'half empty' position were saying 'we told you so', an unexpectedly warm Arctic summer limited the potential damage (World Resources Institute 1993a). The same phrase then issued from the other side. The lesson is surely that we are moving beyond accepting simplistic divisions, to the recognition of the need to understand the dynamic interplays between social, economic and natural environments.

There are efforts in train to review all three of the world's natural resource bases, the social, the economic and the environmental, in relation to one another, after the style of Figure 1.1. United Nations Human Development indices are being calculated for access to environmental as well as to economic resources (UNDP 1993). OECD is working on environmental indicators which can be related to their economic reports; and on the environmental aspects of living conditions (OECD 1993a,b,c). The World Bank reports have begun to consider human and environmental health as important outcomes of economic development (World Bank 1993). Acceptance is increasing that, since the primary goals of most of the world's nations – sustainable development, social justice and supportive human environments – are highly interdependent, they need to be integrated in that nation's policy decisions (Brown 1994). In Figure 1.1, sustainable development is represented as a balance struck between environmental and economic management (Z); social justice as the result of an equitable balance between the production and distribution of resources (X); and supportive human environments as the outcome of social decisions which respect the physical and biological environment (Y). Polarised positions between any of the joint parties would destroy the opportunity for achieving any one of the shared goals.

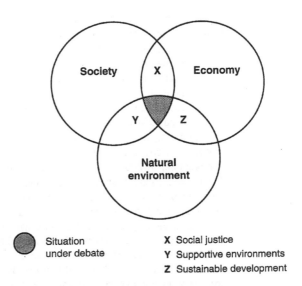

<table>
<tr><td></td><td>Situation
under debate</td><td>X</td><td>Social justice</td></tr>
<tr><td></td><td></td><td>Y</td><td>Supportive environments</td></tr>
<tr><td></td><td></td><td>Z</td><td>Sustainable development</td></tr>
</table>

Figure 1.1 Holosphere of integrated decision-making
(Brown 1994, after Labonte 1993)

Strong dichotomies are still maintained, however. The debate between proponents of conservation and proponents of development, those who believe the
world is going to the dogs, and those who believe there are no worries, constantly
echoes back and forth between the world's regional, national and local scales.
At the global–regional scale, 'North' and 'South' continue to be considered as
having opposing interests in the development of economic and of environmental
resources, respectively. 'East' and 'West' continue to be represented as oppo-
sites in social and cultural matters. Yet Eastern Europe and Russia have
economies closer to the resource-degraded 'South' than to the rest of the
high-production northern hemisphere. The resource management of the Asian
economic tigers, Japan, Korea and Singapore is closer to that of the industrialised West than to the traditional East. Canada and Australia, on opposite sides
of the equator, are each others' closest national counterparts with regard to the
management of their social, economic and environmental resources.

At the national–regional scale, urban and rural concerns are still routinely
treated as if they can be separated. Cities are increasingly being managed as
corporate environments, often without advice from, or liaison with, the environments supplying their life-support systems such as food, water, air and raw
materials – and, most important, their recreational breathing space. As rural

17

areas are managed as mines of natural resources for the denser populated areas, increasing soil degradation, water contamination and air pollution puts these resources in increasing jeopardy.

At national and provincial scales, departments compete for management resources and for the ears of politicians, leaving departments of industry and of environment on collision course. The fresh resources finding their way into departments of environmental management in most modern governments are often dissipated in defending their right to exist. National lobby groups polarise the arguments, since they consider that this will strengthen their case. The few examples of cooperative problem solving have been dramatically successful. For example, the 2000 cooperative community groups established as part of the Australian National Landcare Program to regenerate environmental resources, are a joint initiative of national farmers' and conservationist lobby groups (DPIE 1994). The Ontario economic development agencies, the Ontario Public Health and conservation organisations combined with industry to resolve the management of Ontario's toxic wastes (Hancock 1991). Holland's 'whole country' approach to sustainable development, the National Environmental Policy Plan-Plus, links government, industry, agriculture and households, who all contribute resources to a cross-party long-term environmental strategy (Carley and Christie 1992).

But it is the social mechanisms at the local-regional level that are most frequently called upon to balance the supposed opposition between economic and environmental resource priorities. Local residents remember how their environment was, and are well aware how it is now, either better or worse. They are also the most immediately affected by economic upturns and downturns that may take national and international interests years to register. Local representatives meet their electors in the street, and are themselves affected by the same economic and environmental conditions. The groundswell for change has always arisen from the direct experience at the local scale. The 1980s 'structural adjustment' programmes for economic reform by the World Bank were strongly criticised from the start at the local level, for breaking down community support mechanisms and long-standing customs of natural resource husbandry (Shiva 1983). It was the end of the decade before the social and natural resource balance sheets showed the full disastrous costs of the programme: lowered life expectancy for women and children, and severely degraded natural resource (Commonwealth Secretariat 1989). World Bank programmes now include social and environmental considerations. The perennial seesaw between the driving force of the resource economy and the requirements for continuity of the natural environment must finally be balanced by the social mechanisms which have evolved to resolve conflict and uncertainty in all cultures (Figure 1.2).

The global future is constructed from the patterns of decisions linking the environmental management decisions made at the global and the local scales.

Chapter 28 of Agenda 21 (the United Nations blueprint for environmental management for the 21st century) stresses that local authorities are the point where recommendations made in all the other chapters have to be implemented (Agenda 21 1992). The various global consortia of business, government and non-government organisations (eg the international boards, the G7, the United Nations and the international aid agencies) act on events at the local scale. The outcome of all these deliberations in the global balance sheet lead to a conclusion that the world is half doomed or half saved, according to the particular judge. The actual decision-making mechanisms flow on to the ballot box, the multinational board, the law, ministerial responsibility, community action or even war. Most of these systems depend on representing the options in oppositional terms, not consensus or, better still, mutual support. The climate of the times is 'You win, I lose'; 'More of this for you will mean less of that for me'.

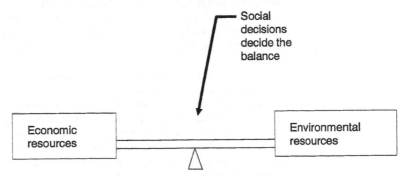

Figure 1.2 Balancing the market economy and the natural environment

The adversarial system of decision making and of conflict management is entrenched in many western, and also many non-western legal, political and business worlds. Parliaments are divided into government and opposition, law courts into prosecution and defence, businesses into allies and competitors. Yet the achievement of sustainable development, that is, a long-term balance between global natural and economic resources, depends on the constructive management of the interactions between the two. Environmental management is as much a matter of human management capacity as it is of technical understanding of the natural environment or the market economy. Yet how to manage when any two different groups look at the same global resource, the planet, and one group argues that it is still full of development resources, and the other that we have already used up more than our share?

The Private Interest and the Public Good

One particularly simplistic dichotomy is used as the justification for arguing against even the possibility of resolving the tensions between the management of the economy and the environment. There has been a long held belief in the supposed permanent opposition of the interests of the individual and those of long-term management of natural resources. This is in marked contrast to the millenia of successful human societies coexisting with their natural environment, in a myriad different cultural systems. The case for a dichotomy between the private interest and the public good in the current environmental debates was put as long ago as 1968, when Garrett Hardin proposed the inevitability of a 'Tragedy of the Commons' (Hardin 1968). He argued that the medieval feudal commons, on which every local resident could graze their own animals indefinitely, is a dangerously misleading model for the present era. The ideal of the commons rests on every individual meeting only his own needs, and all individuals mutually consenting not to place excess pressure on the environmental resource. Hardin argues that those options have gone, if indeed they ever existed.

Hardin quite correctly points out that natural resources are renewable only up to a certain level of harvesting, after which the system breaks down completely. He further argues, as do other market economists, that human beings are innately selfish and will inevitably overstock the commons for their individual gain. Hardin's proposition appears to have been confirmed with the dramatic loss in the diversity of biological species through overuse or removal of the environments on which they depend. Another example would be the dramatic decline of fish populations in virtually all of the world's oceans as a result of overfishing. Conversely, it is possible to conclude that it was a political choice, not the innate nature of human beings, which led to this loss of environmental capital. This view would argue that it is the privatisation of land and reliance on the market economy which removes the option of maintaining the environment for the common good.

Water in some countries, and air in all countries, are natural resources which are still managed in the form of a 'commons'. Nobody owns them; but everybody uses them. Even these 'free' goods are disappearing. Acid rain from large industries killed forests across national borders, producing proposals for new international rules for safeguarding at least that component of the environmental commons (Sargent 1992). Canada suffers from the drift from the United States over the Great Lakes, just as Norway does from German industrial complexes; but there are no means to enforce the protection of those supra-national commons.

Yet until the industrial era was well established, all stable cultures lived in a social–ecological partnership which controlled their particular version of the commons. The actual form of the balance between the availability of the natural resources and the individuals' need for resources for their own well-being differed according to time and place, food sources and shelter materials. The rules for maintaining each particular commons might differ, but the arrangements exist in cultures richly endowed with environmental resources as well as those where they are marginal.

In Bali, where the volcanic soils are so fertile that three crops of rice can be produced a year in an equable climate, food and shelter needs are satisfied with very little claim on environmental resources. Centuries of environmental management have led to highly efficient rice-growing terraces. The unexpended energies of the people are spent on social activities related to the communal societies which maintain the terraces, rather than in raising production levels for sale, or fighting to enlarge territories. The 'commons', that is, the rice terraces, continue to be communally maintained, the communality supported in turn by the time invested in elaborate theatre and dance. Canadian Indians identified different salmon species as part of their own ancestry, and treated them with equivalent respect. In spite of the great abundance in pre-colonial times, the salmon were taken only for necessary levels of food, and so the environmental resources were conserved for tens of thousands of years. In 1993, these are now only 20 per cent of their numbers one century ago (Dorcey 1992). Australian Aboriginal peoples are born into a 'dreaming' which makes them brothers and sisters of a local species. They are thereby bound to preserve the flannel flower, or the *echidna* as if it were their own future – which of course it once was. The Inuit people had strict dietary rules about who could eat what part of a successful kill, ensuring vitamin supplements for pregnant women and kilojoules for the hunters – and the maximum effective use of each animal resource. The cultural rules and the kinship systems act as a mechanism for ensuring social equity, social cohesion and managing the commons at the same time; ensuring the continuity of the biosphere on all counts.

The industrialised world is evolving similar ethical precedents and social rules for lasting human/environment relationships: ones which meet the needs of our own time. The great demand of our era is to manage the way we use technology. With all its enormous benefits, its by-products have been able to puncture the earth's protective ozone layer in one generation and change the global climate within three. As we face the irreplaceable loss of fertile soils, the increase in the still insoluble problem of accumulating radioactive wastes, and the pressure of the newly-industrialised countries for the same level of resource use as the industrialised countries have enjoyed, the need for management systems which can deal with these issues becomes inescapable.

It is becoming clear that those management systems will be marked by several key characteristics. They will go beyond the oppositional, black-and-white thinking towards a collaborative, lateral-thinking, systems-oriented approach. They will need methods of involving all the players, both the current power holders and the emerging ones (including the next generation). And they will need to harness the myriad individual decisions which directly affect the state of local and global environmental resources, in some sort of system capable of protecting the global and local resources.

The Chicken or the Egg?

Just as it is the case that there is no inherent opposition between environmental and economic development, but a strong interdependence; and not necessarily a conflict between the private and the public interest, but strong mutual benefit; so, properly managed, there is no long-term conflict of interests between the global and the local decision making. Even the 'Not In My Back Yard' (NIMBY) syndrome can act as global environmental protection if absolutely nobody anywhere will carry the environmental risk. High-temperature incinerators were hawked around the entire continent of Australia in the late 1980s with no takers. The outcome has been a comprehensive inquiry and a management strategy for intractable wastes which handles them under individual categories, one of the options being waste reduction (Selinger 1992).

In the short term, the immediate local self-interest in a development project may destroy a national or even a global, scarce resource. The Florida everglades and the Phan Pra estuary in Thailand are examples where the effects of local building have had far-reaching ecological consequences in climate changes, flooding and loss of agricultural potential. At least as dangerous as NIMBY to environmental futures are the cases where multinational decisions override local knowledge and identification with place. One example is the continuation of large dam-building projects decades after such major works have been shown to cause massive social disruption, changes in patterns of disease and salinisation, and therefore of doubtful long-term economic benefit, as well as short-term viability due to silting. Twenty-five of the major rivers in North America are now at risk from this cause (World Resources Institute 1993b).

Local or global decisions can be equally disastrous when made in isolation. They must be taken in concert if they are to last or have a successful outcome. Decades of local initiative, however intensive or worthy of global recognition, can be wiped out in a second if there is no integration between likely natural events, and national and multinational actions. For instance, in the 1930s, the local initiatives which were to turn middle America into the world's bread-basket, under the mass production methods which serviced world markets, turned it instead into a dustbowl. In 1992 Indian forests which had been protected by local Chipko women were decimated by one huge landslide. An earthquake dislodged thousands of tonnes of soil from hillsides left denuded through a combination of development exploitation and local poverty. The work of both international aid and local perseverance was gone in a flash (Shiva 1989).

On the other hand, where local support and global interest are one, the long-term interests of both are well served. It was fierce local opposition to nuclear reactors that held up the construction of nuclear power stations long enough for their economic value to be properly evaluated. Few new stations are currently being commissioned in any country except France and Indonesia, thus slowing the accumulation of radioactive waste (Brown 1992). It was the local reaction to the nuclear accident at Three-Mile Island that led to new safety regimes in the nuclear industry everywhere (Keller 1989). It was the rejection by local residents of the expansion of waste disposal plants, from new rubbish tips to high temperature incinerators, which led to commercially successful recycling projects by local authorities in most developed countries.

Local individual consumer choices in a range of western countries has changed national policies and major industry practices. Consumer rejection of tinned tuna caught by drift nets which also trapped dolphin and stripped ecosystems led to bans on drift-net fishing by Pacific countries. Rejection of coats made from wild animal skins (such as seals) or trapped animals (eg foxes) has decimated the fur trade – and the rate of loss of species. In Australia, the willingness of residents to pay more for inland sewerage disposal rather than continue discharges polluting the ocean has led to advances in tree culture using recycled sewerage (Syme 1994).

The United Nations programme, Local Agenda 21, provides the clearest vision of the global nature of the sum of local activities. In the final assessment, the surface of the planet is managed at the local level. Edict, regulation and enforcement of national environmental management methods have never worked without support and advocacy from the local level, from the failure of collectivisation of agriculture in Communist China and the then Soviet Union to the currently unsolved dilemma of transport overload in developed and developing economies. The interactions between economic and environmental priorities are not resolvable by a matter of majority rule, but of joint problem-solving. The guidelines for Local Agenda 21 management plans recognise that it is at the local scale that global initiatives stand or fall. Each Local Agenda 21 management plan is asked to reproduce a locally applicable version of each of the forty chapters of the full Agenda 21.

Any discussion of who has the greatest responsibility for environmental protection, those who live there or those who hold a broader global view, has the same unreal air as does any discussion of the classic aphorism, 'what comes first, the chicken or the egg?'. Neither can continue without the other.

23

Beyond Oppositions

The risk of damaging the environment, and the opportunity for enhancing it, are inherent in any given case of environmental management. The actual state of the environment is, or at least can be, known. Whether it needs to get better or can afford to get worse is a political decision. The evidence on which to make those decisions an advantage to all parties concerned is becoming more and more readily available. As we have noted, more and more government and non-government agencies are now monitoring local and global environments (OECD 1993, World Bank 1993, UNDP 1993). Avoiding the risks and making the most of the opportunities is the job of management.

In traditional environmental management, the options chosen will differ considerably, according to:

- the personal orientation of the manager – optimist or pessimist;
- the professional training base of the technical advisers – social or technological, economic or biological;
- the political priorities of the organisations involved – orientation towards individual rights or the public rights; and
- the scale of operations – local, or global.

We cannot afford to have the options for the future arbitrarily reduced by accepting such false dichotomies. There needs to be a new style of environmental management: one which is inclusive rather than exclusive; which bridges differences rather than accentuates them; and which includes consideration of both risks and opportunities.

A wide range of community and national interests are now asking central questions which overlap with the responsibilities of professional environmental managers. The answers are not all obvious; indeed many of the current school of professional managers would regard the questions as unanswerable. Nevertheless, public concern is making it more and more pressing that they be answered. If one summed up the questions, they might run like this:

- **Ecological unknowns**
  What strategies do we have to ensure we live in an ecological balance which does not threaten our future?
- **Demographic certainties**
  What do demographic projections indicate will be the social pressures helping and impeding the implementation of strategies for ecological sustainability?

- **Economic probabilities**
  What trends in local and global economies will maximise and which will minimise, degradation of environmental resources?
- **An informed public**
  Given the powerful and ever-increasing influence of the mass media on social behaviour, what is the relationship between public information and environmental management?
- **Local self-reliance**
  Local self-reliance is a powerful motivating force and source of action. Can environmental managers work directly with communities, as well as services, to implement political and scientific recommendations?
- **Environmental public policy**
  Public policy is about organising the future; about generating change; about maximising opportunities. What are the elements of public policy necessary to achieve sustainable development?

These questions issue a set of challenges to environmental management. In order to respond to them, environmental management is changing. The questions remain:

Which direction should environmental management be taking?
What will it manage?
How will it be managed?
Who will manage?

References and Further Reading

Boyden, S, Dovers, S and Shirlow, M (1992) *Our Biosphere Under Threat* Oxford University Press, Melbourne

Brown, L R (1989–93) *State of the World 1989–93* (Worldwatch Institute) Earthscan, London

Brown, L R (1993) *Vital Signs* Earthscan, London

Brown, L R (1991) The new world order in *State of the World 1991* (Worldwatch Institute) Earthscan, London

Brown, V A (1994) *Acting globally: the key role of Local Government in environmental management* Department of the Environment, Sport and Territories, Canberra

Carley, M, Christie, I (1992) *Managing sustainable development* Earthscan, London

Commission for Sustainable Development (1993) *Agenda 21: Work program 1993-98 thematic clusters* United Nations, New York

Commonwealth Secretariat (1989) *Engendering structural adjustment for the 1990s* Report of a Commonwealth expert group on women and structural adjustment, Commonwealth Secretariat, London

Department of Primary Industry and Energy (1994) *The National Landcare Program*

Dorcey, A H J (1991) Sustaining the Greater Fraser River Basin, in Dorcey, A H J and Griggs, J R G (eds) *Water in sustainable development* Westwater Research Centre, Vancouver, pp 269–82

Hancock, T (1990) *Community, Economy, Environment* Briefing paper for Canadian Public Health Association Conference on Health and Environment, Ottawa

Hardin, G (1969) The tragedy of the commons *Science*, 162, pp 1243–8

IUCN/UNEP/WWF (1980) *World conservation strategy: living resource conservation for sustainable development* IUCN/UNEP/WWF, Gland, Switzerland

IUCN/UNEP/WWF (1991) *Caring for the earth: a strategy for sustainable living* Earthscan, London

Mathews, J T (1990) Environment, development and security *Bulletin of American Academy of Arts and Sciences,* vol xliii, pp 10–26

National Environmental Policy Plan (1989) *To choose or to lose* SDU, Uitgeverij Gravenage, Netherlands

Holmberg, J (ed) (1992) *Policies for a small planet: from the international institute for environment and development* Earthscan, London

OECD (1993a) Terms of reference, project group on the ecological city (mimeo), Environment Directorate, OECD Paris

OECD (1993b) Environmental indicators: a preliminary analysis (mimeo), Environment Directorate, OECD Paris

OECD (1993c) Urban indicators: analysis of existing literature and main issues (mimeo) Group on Urban Affairs, OECD Paris

Keller, K H (1989) Science and technology *Sea-changes: American foreign policy in a world transformed* Council on Foreign Relations, Washington

Labonte, R (1993) A holosphere of healthy and sustainable communities *Australian Journal of Public Health,* vol 171, pp 4–12

Meadows, D H, Meadows, D L, Randers, J (1992) *Beyond the Limits: global collapse or sustainable future?* Earthscan, London, p xv

Pearce, F (1988) A damned fine mess *New Scientist,* April

Sargent, C, Bass, S (1992) The future shape of forests, in Holmberg, J (ed) *Policies for a small planet: from the international institute for environment and development* Earthscan, London

Selinger, B (1992) *A cleaner Australia: report of the independent panel on intractable waste* Independent Panel on Intractable Waste, Paddington, Sydney

Shiva, V (1989) *Staying alive, women, ecology, development* Zed Books, London

Shiva, V, Sharatchandra, H C, Bandyopadhay, J (1983) The challenge of social forestry, in Fernanades, W and Kulkarni, S (eds), *Towards a new forest policy* Indian Social Institute, New Delhi

Syme, G (1994) *Report, Australian Research Centre for Water in Society* CSIRO, Perth

United Nations Development Program (1992) *Human Development Report 1992* Oxford University Press, Oxford

United Nations Conference on Environment and Development (1992) *Agenda 21: Chapter 28 Local Authorities initiatives in support of Agenda 21,* in Report from UNCED, Brazil

World Bank (1991) *The challenge of development: world development report 1991* Oxford University Press, Oxford

World Bank (1993) *Development and Health* Oxford University Press, Oxford

World Bank (1991) *The challenge of development: world development report 1991* Oxford University Press, Oxford

World Resources Institute (1993a) *Environmental almanac: restoring water habitats* Houghton Mifflin, Boston pp 46–8

World Resources Institute (1993b) *Environmental almanac: ozone depletion* Houghton Mifflin, Boston pp 303–12

World Commission on Environment and Development (WCED) (1987) *Our common future* Oxford University Press, Oxford

2 The Challenge of Changing Environmental Management

Why is Environmental Management a Personal and Professional Challenge?

Practical Challenges

The Western cultural set of identifying the elements of a dilemma as in opposition to one another serves environmental management very badly. This is true at whatever level we discuss: local, national or global. Until the mid-eighties, the management of the industrialised world's natural resources was dominated by single profession organisations. Water boards were advised by engineers, mining councils by geologists and so on. The training of the wide range of professions involved in environmental management was almost exclusively technical and specialised. The training of professions perceived as having no direct connection with environmental management included no mention of environmental issues, even as these became all-pervading. This put individuals in a poor position to cope with multi-faceted issues, and increased the likelihood of a them-and-us mentality.

In the 1990s, scientific, legal, financial and political organisations remain divided on the proper methods for managing environmental issues – their nature, the ways of dealing with them, and the relative importance of conservation and development. The debate over the Greenhouse Effect and how to respond to it is a prime example of conflicting opinions barring the way to constructive solutions. The re-establishment of wind power as a fuel source in Holland and the halt in the decline of the Canadian salmon industry are examples of multi-interest, multi-scale and multi-disciplinary cooperation in solving environmental issues.

To expect this type of cooperative solution to be easy would be naive. The custom of adopting expert advice on each separate facet of an issue is so deep-seated as to be the cultural norm. To change the system of unchallenged supremacy in one area is, in the first instance, to increase the degree of conflict. When different groups work together on a particular issue, they bring different languages, terms of reference, professional loyalties and basic understanding of the problem. What different professions see as the problem varies enormously. In issues as different as the salinatisation of the land in the Australian Murray-Darling Basin, the drastic decline in salmon stocks in Canada and the effect of urbanisation on Bangkok, the early attempts at intervention by single-profession experts did nothing to resolve the issues (see Chapters 6, 7 and 8). The tensions generated at the time escalated the conflicts dramatically and built higher barriers between the parties attempting to resolve them. In each case, it has been the closeness of the issue to all members of the affected communities which has led to a renewed attempt at solution.

To change the mode of management from fragmentation and confrontation to coordination and cooperation is no simple step. It amounts to no less than a new culture of environmental management, requiring terms of reference which integrate social, economic and environmental concerns; an inter-professional loyalty; combined decision-making; and a composite understanding of the problems and their solutions.

This dramatic shift is not a pipe dream. There is a worldwide trend towards including all the players in the management of any given environmental region or issue. Conflict resolution and management skills, open communication, coordinating structures and collaborative techniques have entered mainstream resource management. Both Agenda 21 of the United Nations and *The Way Forward* of the International Union for the Conservation of Nature incorporate conflict resolution mechanisms in their central structures.

At the national scale, Canada and Norway have adopted 'Round Tables' to coordinate the multiple interests in environmental professional activities. Scandinavia, Canada, Australia and the European Union have national sustainable development strategies, each with some form of multi-party collaboration. Integrated catchment management is an increasingly common basis for environmental management, providing links between political and geographic scales. At the local scale there are numerous valuable examples of a new culture of cooperation, from the combination of landholders and conservation interests in the Australian National Landcare programme to the consortium of senior Buddhist monks, junior academics, local residents and government officials re-examining the role of transport in Bangkok to the management committee for the Pacific salmon including indigenous fishers, ecologists, major fishing industries and national and provincial government agencies in British Columbia.

Alliances within professions or within government or non-government agencies have evolved to establish professional territories and improve specialist expertise. The alliances can therefore actually be detrimental to multi-skilled management of environmental matters. The membership of most professions associated with environmental management are split between conservation and development, and between private and public interest practitioners. Local government interests focus on the local aspects of global concerns, as with coastal authorities and the risks of coastal flooding from climate change. National agencies supervise local implementation of national responsibilities, as with United Nations' national programmes for Local Agenda 21. The common interests are the connecting theme, rather than the professional or government positions.

The risk of damaging the environment, and the opportunity for enhancing it, are not pre-determined for any given environmental management issue. The risk and the opportunity will be the outcomes of the management's decisions. In Chapter 1 we argued that the options may be based on:

- the personal orientation of the manager – optimist or pessimist;
- the professional training base of the technical advisers – technological or biological;
- the political priorities of the organisations involved – orientation towards individual or public rights; and
- the scale of operations – local, national or global.

All those involved in environmental issues – lawyers, politicians, economists, planners, scientists, community representatives and more – increasingly need conflict management techniques and the personal skills to apply them. Such skills are not only rarely included in their original training but may be in contrast to the way they are expected to operate in other forums. There is a need for a shared set of simple principles for managing environmental issues, straightforward enough to be applicable to diverse situations and by diverse interests. The principles must be founded on terms of reference for environmental management which include the following:

- recognising that managing the environment means managing social change, reconciling socially-constructed oppositions between:
    - the private and the public good
    - government and non-government interests
    - established professional positions
    - the present and the future
    - the global and the local scale;

- reconciling oppositional positions, making conflict management an integral part of environmental management, incorporating all the players, whether mainstream or marginalised;
- ensuring that managing conflict is not a destructive but a constructive process, capable of minimising the risks and maximising the opportunities for future sustainable development.

In all countries, the environment urgently needs good management. For example, more than 50 per cent of Australia's land mass requires remedial treatment. Over half the national water consumption goes to irrigate the one river basin, with the result being increasing levels of salination. The predicted changes in climate due to a global temperature rise will have a severe effect on a country which is already subject to severe droughts and floods as part of its normal weather cycles. Australia's energy use per head has doubled in the past 25 years and is now higher than any other industrialised country except the United States. City wastes are increasing at an average of nine per cent every year, and waste disposal is now a matter of acute community concern.

On the positive side, Australia has large amounts of unutilised space, vast reserves of natural resources, and a high level of individual health. On OECD scales of success for industrialised countries, Australia is ranked fifteenth in Gross National Product, and seventh in quality of life. Australians have provided leadership in both the development and conservation of natural resources. From Farrer's 'rust-resistant' wheat in the last century to CSIRO's 'gene shears' in this, Australia's biological research has contributed to a new agriculture. Yet it is now the country with the world's greatest proportion of degraded land.

Caught between these national strengths and weaknesses and the overall increase in public concern, environmental decision-making becomes a greater challenge at every stage. Economic debates on the competing uses of environmental resources have become angrier and more urgent. Community awareness of pressures on the environment has heightened. The establishment of the first Australian department of the environment and the first national government intervention to preserve an ecosystem (the largest sand island in the world, Fraser Island) in the 1970s changed the social and political context irreversibly. Since then, there have been major changes in both social and natural environments. Only 15 per cent of Australia's population remain in rural areas, where 30 per cent lived at the turn of the century. It is predicted that this level will be reduced by half again in the next five years. Fifteen thousand hectares of agricultural land are lost every year to the spread of cities. The national expertise in agriculture, soil conservation and environmental management has improved greatly at a technical level. Courses have been mounted in environmental management at all of Australia's sixty or so tertiary institutions.

Unfortunately, these are usually segregated within Biology faculties; and while aspects of environmental management are creeping into engineering, health and policy courses, these rarely include management of human as well as natural resources. The Prime Minister of Australia, whoever he may be at the time, has taken to making annual policy statements on the environment. The budget allocation for national environmental management does not increase, however. In general, the Australian situation is closely paralleled in North America and Europe. Environmental consciousness has risen; advances have been made.

In national environmental policy; technical expertise continually advances; but the resolution of environmental disputes and of environmental problems lags far behind.

Moving north to Thailand, the dilemmas are much the same; they merely take a different form. A tropical country, Thailand, unlike Australia, has an assured monsoonal water supply. The issue here is the control of that water: assurance of a predictable supply to rice-growing areas; control of the rapid urban sprawl of Bangkok across the Chao Phraya River delta; and the protection of groundwater which acts as an unregulated water supply to the eight million inhabitants of a city which has expanded without much time or attention to give to infrastructure. In a few short years, as Bangkok acted as a supply base for the Western forces in the war in Vietnam, the water canals which had given Bangkok the name of Venice of the East were filled in, and replaced by concrete highways with underground sewers. As if that was not sufficient environmental disruption, impeding the natural flushing of the river system and the natural flood control of the delta, there are now 2 million vehicles on those highways, and on the unpaved roads which are ramifying out over the rice-fields. With the current rate of increase of 25 per cent a year, there will be 1.8 cars for every household by the year 2000. The traffic surface area is 2.5 per cent of the city; the recognised standard for Western cities is about 25 per cent. The continual traffic gridlocks on this small road base are now proverbial.

The management system for these rapid changes is a highly centralised government, with few regulatory instruments; and those there are have well-established loopholes. For instance, 10 per cent of buildings in any area do not have to conform to regulations. Central government also acts as local government, so that national and local planning permission flow through the same channels. Government departments and agencies are strictly hierarchical and strongly territorial. Employees do not move from one to the other and are expected to be loyal to their colleagues; so there is little opportunity to consult with fellow professionals in another area. On the other hand, both residential districts and extended families are highly cooperative, so that community action on local issues is more likely than official action. So while environmental concern may stem from the community, as in Australia, it is more likely to reach senior decision-making levels through informal, rather than formal channels.

Trends towards urbanisation, increased petrol consumption, degraded agricultural land, uncontrolled flooding, reduction of natural resources and population increase have sweeping implications for all concerned with environmental management. This includes almost everybody: options for the future use of resources are controversial for industry, community and government alike.

Formal avenues for resolving environmental debates are emerging at every scale. They range from international conventions, commissions and agencies; national land and environment courts, special committees of inquiry, multi-interest round tables and national lobby groups; to local committees, councils and action groups on special issues. Most can be identified as falling under one of six types for resolving environmental debates:

1   **Legal-judicial** mechanisms invoking the use of the courts and special tribunals;
2   **Political** avenues, including three different levels of elected representative (local, State and Federal), and the possibility of a referendum;
3   the **market economy** through the pressures of supply and demand, adjusted by pricing and taxes;
4   **Bureaucratic-administrative** procedures through impact assessment, regulation, licensing and statutory agencies; and
5   **Scientific** inquiries evaluating and coordinating technical evidence.
6   **Public participation**, ranging from consultation with a local community group through to partnership in a national programme.

Continuing major changes in the use and regulation of natural resources leave environmental managers with no choice but to be managers of environmental change. Similar changes are operating at the personal and professional level as well. Change is occurring within the ranks of environmental management and all its associated occupations. The tasks of an environmental manager now range from coordinating technical interpretations of the latest Greenhouse forecast to chairing public dispute meetings. The label of environmental manager now applies to almost all occupations, political levels and government agencies.

One of the more radical changes found in most western countries is that policy and protocols on natural resource management now refer to some measure of public involvement, at least in principle. In practice, public involvement means many things to many people. Examples range from government and industry-sponsored Save Water and Green Choice campaigns, through the formal public inquiries to the grass-roots self-management of Catchment Management Committees. Within this broad spectrum a myriad options, from the conventional to the creative, are chosen by the protagonists to fit the appropriate situation.

Five distinct uses of the term public participation can be identified in the literature, with the extent of public involvement increasing:

1   a means of convincing the public of the value of decisions already taken – more honestly described as public education;
2   an activity undertaken at the discretion of decision-makers, if the situation warrants it and resources are available – better called public contribution;
3   a contribution to project management, in which public opinion is canvassed at certain stages of the process – public consultation;
4   a method of conflict management where there are a range of apparently incompatible positions, as in reviews and public inquiries; or
5   an integral part of the decision-making process, in which the relevant publics are regarded as partners in the policy development, management and monitoring of a natural resource, that is, public partnership in the enterprise.

Water issues in particular have a strong track record of public participation, involving the public interest in the decisions of any of the management sectors responsible for a project. In the United States, the Army Corps of Engineers has pioneered conflict resolution and mediation between the different interest groups, based on the Corps' responsibilities for water courses and catchments, which are federally owned.

Modern Environmental Impact Assessment procedures accept that the public:

- have a legitimate interest in any environmental change;
- have skills to contribute to evaluating the change;
- have resources to supply to implementing the outcome; and
- provide an effective contribution to government decision making on a project (Atkins 1984; Brown 1986).

Public participation in impact assessment at each scale requires a different alignment between public decision-making on the one hand, and public interest and lobby organisations on the other. The aim of effective public participation is to ensure that this alignment is managed so that potential conflicts are redirected towards cooperative and constructive partnerships; local interests are taken into account; and policy requirements are met. The ideal is very far from the actual in most consultation processes; but nevertheless the ideal of public participation has entered the environmental management repertoire of professional practice.

There is a wide range of possible avenues for involving the public:

- **Formal:**

    - legislative requirements, eg environmental health regulations
    - voting
    - lobby groups
    - round table discussions
    - arbitration
    - data collection, eg state-of-the-environment reporting
    - strategic planning
    - open public forums, widely advertised
    - commissions of inquiry such as royal commissions
    - regional conservation strategies
    - regional development corporations
    - Local Agenda 21

- **Informal:**

    - networking, as in person-to-person contacts
    - teleconferencing, by phone or electronic mail
    - key informant interviews
    - phone-ins and write-ins
    - vision workshops where an ideal future region is designed
    - focus groups
    - whistle-blowing
    - gossiping
    - tea and lunch breaks
    - hotlines
    - neighbourhood centres
    - special events

The new ground rules for environmental management are markedly different from the old. Tables 2.1 and 2.2 outline the old and new rules. Traditionally, environmental management was highly controlled, technical and fragmented. Water, air and soil were managed as separate and unlimited resources. Water resources were considered only at each point of use; as a dam to be built, or a city to be sewered. Today, catchment area, storage, river, human settlements and estuaries are each accepted as being part of integrated catchment management.

Table 2.1 Environmental management: the old rules

- Water, air and soil are inexhaustible resources.
- Environments can be managed in sections.
- Social values do not change.
- Environmental management is a technical matter.
- Technology can make anything possible.

Table 2.2 Environmental management: the new rules

- Water, air and soil are finite resources.
- All parts of an ecosystem are interrelated.
- Social values are changing.
- Environmental management is a people matter.
- Use of technology has limits and costs.

We are witnessing the evolution of an integrated and more cautious management. Natural resources are accepted as being limited. Natural resources are recognised as being interrelated within self-balancing natural ecosystems. Social priorities are changing with respect to natural resource allocation. Environmental management has become the management of people, environment and technology combined. No single one of these provides all the answers.

The move from a loose collaboration of technical experts to a consultative precautionary management team has taken place in little more than a decade. This rate of change would generate internal tensions within any workforce. The resulting tensions lead to the conflicts which are inevitable in social and environmental change. For the foreseeable future environmental management will be management of conflict and change; both internally within the management team and externally in coordinating environmental resources.

Conflict in itself is not necessarily a bad thing. It signals the recognition of a need to change. Change is the bridge between the old and the new, but it also inevitably brings conflict between the two.

In Chinese, two characters are used to represent the word 'crisis'. One specifies risk and the other opportunity. This definition can equally well be applied to conflict. Conflict may result in disaster, or provide a highly useful avenue for change, or something of both. For a positive outcome, resolution of conflict requires careful and skilled handling and a major revision of how our society regards conflict itself. Table 2.3 lists five characteristics of conflict which are often misinterpreted by participants and onlookers of the conflict alike.

Table 2.3   Conflict and environmental management

- **Conflict is an inevitable part of change**
  It is not the result of personal failure,
  nor the failure of the system.
- **Conflict is a step in the solution to a problem**
  It signals the opening up of debate.
- **Conflict is shared**
  It is not the sole responsibility of any group or set of interests.
- **Conflict is a process**
  It is not a result, a barrier or an excuse.
- **Conflict is manageable**
  But management takes time and resources.

A conflict establishes that there is more than one point of view on an issue. Absence of conflict is not always desirable, since it may indicate apathy, ignorance or powerlessness on the part of those affected by change. Each of these conditions can create a greater barrier to environmental sustainability than conflict which is well managed. For example, no-one listened to the isolated residents of the asbestos mining town of Wittenoon in West Australia for three decades after lung cancer risks were confirmed. Now, the area is uninhabitable and every State Health Authority has a case load from asbestos that will last well into the next century. The shift from the old rules of environmental management to the new have reasons for conflict built in. Changing social values bring confrontation between age groups and between the innovative and the conservative. Limitations to natural resources and to technology mean that choices must be made.

Conflict is part of any process for resolving differences. While conflict is part of the process of managing change, it is not the end result unless allowed to become so. Currently, conflict has become embedded in the process of forest management in all countries. In most cases even embedded conflict is resolved eventually, but rarely without active intervention and consent of all the players involved. The World Heritage listing of the rainforests was a process of mixed consultation and confrontation.

Environmental management involves minimising risk and maximising opportunity in matters of the environment and in matters of management. Diagnosis and reduction of risks to the environment are coupled with the opportunity for enhancing existing environmental resources. The chance to manage other peoples' conflicts provides many opportunities for constructive change.

References and Further Reading

Atkins, R A (1984) 'Comparative analysis of the utility of EIA methods' in Clark et al (eds), *Perspectives on environmental impact assessment* D Reidel Publishing Company, Dordrecht

Axelrod, R (1984) *The evolution of cooperation* Basic Books, New York

Brown, V A (1984) Healthy cities: community co-operation for policy and action *Proceedings, First National Community Health Conference*, Adelaide

Day, D (1991) Resolving conflict in the New South Wales water sector: informality, anomaly and innovation, in Handmer, J W, Dorcey, A J and Smith, D I, *Negotiating water: resolving conflict in Australian water management* Centre for Resource and Environmental Studies, Australian National University, Canberra, pp 168–90

Fisher, R and Ury, G (1987) *Getting to yes: negotiating agreement without giving in* Arrow, New York

Handmer, J W, Dorcey, A J and Smith, D I (1991) *Negotiating water: resolving conflict in Australian water management* Centre for Resource and Environmental Studies, Australian National University, Canberra

Kuhn, T (1970) *The structure of scientific revolutions* Routledge, London

Ury, G (1990) *Getting past no: negotiating with difficult people* Business Books, London

Chapters 6, 7, 8 and 9 of this volume

3   What to Manage?

The five stages of environmental management

In any environmental conflict, *what* has to be managed includes people and projects: a wide range of people holding different points of view, and issues which are likely to be controversial. Environmental management can best be regarded as a set of decision-making stages which link the people and the issues.

Whether the task is management of an issue for a small creek or a large city, the five decision-making steps are the same:

1   **Exploring** the ramifications of issue;
2   **Analysing** the results of the exploration;
3   **Setting goals** for action;
4   **Taking action;** and
5   **Evaluating** the process and its outcome.

Before deciding on the options, the important influences on the particular environment need to be explored. Does the creek need a dam or better conservation of its headwaters? Does the city need a new airport runway, a new airport or less air traffic? Analysis of a wide range of information allows the manager to identify the key players and resources. This in turn allows for realistic goals to be set for practical actions. It is of little use opting for a dam if there is no funding. It is no use deciding to build a runway if it is known to be politically unacceptable. Evaluation of the results of the actions involves consideration of the interests of the key players and the future steps to be taken.

The new ground-rules for managing the environment include economic, ecological, social and technical aspects of every issue. This may make it seem as if everyone on the earth has a legitimate concern in every issue. At one level this is true. However, for practical purposes the above five aspects of any one issue are crucial, as are the three different groups of people involved in the management (Figure 3.1).

Among the players in Figure 3.1 are those who have a stake in the outcome; those who possess important relevant information; and those who will be needed to contribute directly to management. The three categories of player are not mutually exclusive. A local resident has unique information and may be appointed to the management committee. Each expert who advises the project has a stake in its success.

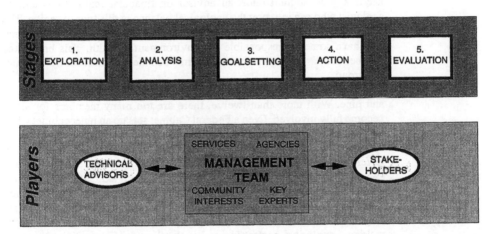

Figure 3.1 Components of environmental management: stages of decision-making and players involved

Given the task of managing, for example, Calico Creek – a small unregulated catchment with a low population in a high rainfall area (subject of the case study in Part 2) – who would be included in the management team? Obviously, the local residents, the local shire and the local town services. The people downstream also have a major interest. There will be regional regulatory agencies, and a national policy for the area. The local produce has consumers in the State capital as well as in outlets in Asia. All these are stakeholders with a legitimate interest. To ignore any one of them would be at peril of arousing later confrontation which could then be difficult to resolve.

Advisers

As well as local knowledge and support, management needs a range of expert advice. For example, that advice may be about soil chemistry, rate of water use, or management strategies. The first step in managing potential conflict is to identify all the necessary advisers. The second is to bring them together in order to coordinate the advice. In Calico Creek, one would include the shire engineer and a representative of each main group of users: agriculture, grazing, domestic supply and town council. The management team itself needs a good administrator, an adviser on financial implications and an ecologist who can interpret the ecological balance in this type of region. The State may have a good soil conservation officer available. The local council will have someone responsible for environmental health. This brings the number of the working group to ten, almost an ideal number for a coordinating group.

A social psychologist would advise a team membership of between seven and nine. With more than twelve, there are too many members for all to become fully involved. Even a Federal Cabinet sets up an inner cabinet when the number of Ministers goes over twelve. If less than seven, the group is too small to share the tasks, and small enough for personality conflicts to develop. Determining the mix between services, agencies, community and experts is one of the essential conflict management skills. Suggestions for doing this are discussed in the following section on how to manage conflict, as well as in the learning modules. The chosen mix will differ at different stages of decision-making. At the learning stages of environmental decision-making – exploring, goalsetting and evaluating (see Figure 3.1) – the more of the players involved the better. At the two task-centred stages alternating with these – analysis and action – it is the players in the smaller management team who count.

Table 3.1 Environmental management: key themes at each stage

Economic issues:	What will it cost?
Ecological issues:	What will it affect?
Social issues:	Whom will it affect?
Management issues:	How can it be done?
Technical issues:	Can it be done?

Common Features of Environmental Issues

All environmental issues have certain features in common. The new management rules emphasise the need to consider the economic, ecological, social and technical aspects of each issue (Table 3.1). Returning to the theoretical example of Calico Creek, the economic issue is not directly related to water management *per se*. It is about whether the area should service bigger markets, build dams and charge water rates; or whether to maintain the quality of life in naturally beautiful surroundings for the existing residents. The political economy of Calico Creek operates on a cooperative rather than competitive basis. Except in times of drought, the water supply is enough for all current water users, so long as those users agree between them to balance their use. Community cooperation and supervision are essential in times of scarcity, but the community-imposed rules have been established well before the potential conflict arises.

The new rules for environmental management lead to new questions of management on each of the five areas: what will it cost; what natural balances may it affect; whom will it affect; how will the decision be implemented; and is it technically feasible? (Table 3.1). These questions must be answered at every stage of the process. In the case study of Calico Creek, the ecologists advise that the district is under good management. It is not overgrazed, use of fertilisers is low and there is little erosion. Economically, there is a very wide range of ownership of resources, with one big property owner, several middle-sized owners, and a few small-holders. A range of reliable markets allow for diversity in production. The area is politically conservative and pro-development, except not necessarily in their own backyard. There has been a community management committee for water allocation for years. This meets mainly in times of drought. Technological questions arise from local pressure for a dam and management needed for a water metering system.

Three further case studies in Part 2 also illustrate the five components of each environmental issue. In Nyah Shire the political economy is an uneasy mix of private enterprise and State subsidy. The products sold off the land have been estimated to earn less than the subsidy. Ecologically, irrigation and deforestation have led to high and increasing salinity. Only salt-tolerant crops survive and downstream pollution is severe. Socially the district has strong Greek, Italian and long-established Anglo-Saxon farming communities. Management is through Federal and State agencies, local community control and local shires.

Technically, there is an argument for an engineering solution. Laser levelling (making all fields geometrically horizontal) would reduce the overall amount of irrigation needed, and prevent the collection of heavy pockets of water. Another proposed solution is to buy current agricultural users off the land, and return to dry-land farming or sell to hobby farmers.

41

Meanwhile, small pockets of land are still unaffected and their residents are outwardly unconcerned.

It is important to distinguish between the five types of question in Table 3.1. In the case study of Bangkok, the rapid industrialisation is bringing economic growth to Thailand at a rate higher than any country in Asia outside Japan and Korea. The rapid increase in national income does not reach the agricultural areas, nor the agricultural labour. The richest 20 per cent of the population receive 55 per cent of the national income; the poorest 20 per cent receive less than 5 per cent. The ecological costs of the move from almost entirely an agricultural, rice-growing nation to manufacture, services and industry are not beyond calculation – but they are not yet being calculated. They certainly include air pollution; although this cannot be evaluated, since only the monthly averaged levels of lead, carbon monoxide and suspended particles are published, smoothing out the peaks and troughs. The reduction and pollution of ground water supplies is traceable from the increase in infectious diseases in certain parts of the city.

The issue of water supply is also a social issue. Who pays and how much are social equity and political issues. The pull of agricultural workers from self-sufficient farms in the hills to social displacement in the city; and the equivalent urban spread over close-in farmland, with displacement of labourers has the joint effect of degrading land, and displacing people. Management we have seen is still evolving, towards a Thai adaptation of the western style bureaucracy. Technical skills are well-represented in the national agencies; and in the many aid programmes and agencies, particularly in the country. The bridge between the skilled and unskilled is as wide as that between income levels. Thailand, however, has had generally high literacy for several generations now; and a high level of traditional farming expertise.

Three sets of players involved in management and five key themes for each management decision together make up *what* environmental management does. The risk of conflict between the players is great; but so are the opportunities for collaboration and cooperation. The range of potential options for solution of each type of problem is even greater. Management of conflict and change is a matter of exploring the wide range of options. Mutual agreement between all the players on the possible range of options is the first step towards a successful outcome.

Social and Economic Implications

So far we have considered the management of the natural environment as the management of (a) the effects of humans on the environment, and (b) the inter-relationships between environmental processes themselves. The effects of social and economic conditions on environmental decision-making are addressed far less frequently, if at all. Models of economic and social change as they affect environmental management are emerging in the literature. A review of the relationship between economic policy and environmental management, *For the common good: redirecting the economy toward community, the environment and a sustainable future* proposes a set of indicators for monitoring the effects of economic decisions on the natural environment (Daly and Cobb 1989). Unfortunately, these proposals fall short of including the monitoring of the social influences on, and effects of, environmental change. Another such attempt, *A blueprint for a green economy* (Pearce et al 1991) offers a series of economic instruments for protecting the environment and industry from excesses of uncontrolled development. The work has been criticised for considering only matters of the market economy, while much of the value of the natural environment and the costs of protecting it lies outside the economic market. A third publication, *The green economy: environment, sustainable development and the politics of the future* takes as its main theme the interactions between social, economic and environmental considerations (Jacobs 1991). An economics of diversity is developing, one which is likely to meet the requirements of being able to contribute to future environmental management, through:

- seeking an economic synthesis between market, state, family, community and environmental resources;
- recognising the rationality in human decision-making which is able to evaluate complex trade offs between risks and opportunities;
- determining the optimum levels of environmental protection within and without the market; and
- placing more emphasis on the goal of environmental sustainability as an economic good in its own right (Brown 1993).

These are hopeful directions, rather than existing tools; but at least they are on their way.

The effect of their environment on human beings has a crucial effect on human decision-making, and so in turn on environmental management. Public Health, the specialised field which has evolved to deal with the effect of the environment on humans, has developed a model for effecting social change towards better health. The model applies equally well to human-environment interactions. Labelled informally the New Public Health, and formally the Ottawa Charter (WHO 1987), the approach provides a comprehensive framework which adds community action and social policy to the traditional elements of environmental management.

43

Public Health has been addressing environmental issues for some time. The Ottawa Charter was being produced in Canada at the same time as the World Commission for Environment and Development were preparing *Our Common Future* on the need for concerted action on environmental issues. The Ottawa Charter 1986 defined public health action as a combination of five interrelated strategies, which can equally be related to environmental management, as follows.

1 **Coordinating policies**: so that housing, transport, environment, education and law enforcement relate the effects of their policies and programmes to each others' – and to the overall well-being of the community and of their natural environment.

2 **Enhancing social and physical environments**: by recognising the need for full and public evaluation of those environments, so citizens can monitor what is happening, and services can set priorities – integrated state-of-the-environment reporting is beginning to fulfil this function.

3 **Strengthening community action**: is a necessary basis for any public health intervention – and also for any lasting environmental protection, since it is not possible to change social goals towards sustainable development unless the mass of the citizens agree.

4 **Developing individual skills**: in safeguarding one's own and one's community's health – including knowledge and advocacy skills for safeguarding a local environment;

5 **Reorientating services and agencies**: to give a stronger emphasis to prevention, is a logical step, since over 80 per cent of contemporary health risks are known to be preventable – all equally true of the precautionary principle in environmental management.

For an environmental issue, say, the Canadian salmon fishing industry, a social change strategy based on the Ottawa Charter would include:

- public policies which incorporated economic, environmental, indigenous peoples, primary industries, science and education portfolios;
- state of the environment reporting, conducted on a bio-regional bias in Canada, in conjunction with Statistics Canada to link the socio-demographic and economic profiles of the Fraser Basin;
- community agencies such as American Indian and recreational fishing groups provided with regulatory and financial support, and included in the management councils, together with industry and scientific advisers;

- individual understanding of the interactions affecting the bio-region promoted through media articles and interviews, and educational materials at all levels; and
- reorientation of the multiple government and non-government agencies responsible for the salmon industry, through establishing collaborative mechanisms rather than parallel chains of command, and agreeing on the precautionary principle as a basic principle for combined action.

The Ottawa Charter permits a full diagnosis of just *what* will need to be managed throughout the five stages of environmental management. It allows us to identify the potential alliances, or equally potential conflicts, which could arise in the process. The next difficult question is, just *how* is this complex system to be managed?

References and Further Reading

Brown, V A (1993) 'Where the buck stops: towards an economics of diversity' in *Proceedings, Conference on environmental economics* Department of the Environment, Sport and Territories, Canberra

Jacobs, M (1991) *The green economy: environment, sustainable development and the politics of the future* Pluto Press, London

Daly, H, Cobb, J (1989) *For the common good: redirecting the economy toward community, the environment and a sustainable future* Beacon, Boston

Pearce, D, Markyanda, A, Barbier, E B (1991) *A blueprint for a green economy* Earthscan, London

Chapter 6: A Heroic Effort: the salmon fishing industry in British Columbia

Chapter 7: Saving a Shire: a salt-affected irrigation district in New South Wales

Chapter 8: Trouble with Traffic: rapid urbanisation in Bangkok

Chapter 9: The Future of Calico Creek: local management of a small water catchment

4 How to Manage?

The role of conflict management in environmental management

Each stage of environmental decision-making presents a different challenge to management. Opening up an issue to public comment presents a challenge to existing interests, who usually prefer things the way they are. The final evaluation stage can pit each player against the others in claiming credit or shelving blame. From the small self-sufficient rural economy to the turbulent industrial city, there is potential for conflict from competing specialist advisers, conflicting environmental interests, and disparate skills within the management team. Consequently, there is every reason to incorporate conflict management into environmental management at each of the decision-making stages.

The management task becomes the identification of the appropriate conflict management strategies and correct tools for each of the five stages of the management process. Figure 4.1 suggests the most appropriate conflict management strategy at each stage. Strategies are selected from the wide range available, and each can augment the other.

Conflict management methods range from war at one extreme to complete isolation and non-involvement at the other. The possible range of suitable methods is indicated in Figure 4.2. At the extreme ends of the management range are those approaches best described as a failure of the system. **Apathy** on the one hand, and **war** on the other are the opposite poles to cooperative management. **Isolation** and **confrontation** are the conflict management techniques applied at these extremes. The use of either by any of the parties makes constructive management difficult. Any strategies falling short of these two extremes will be a step in the right direction.

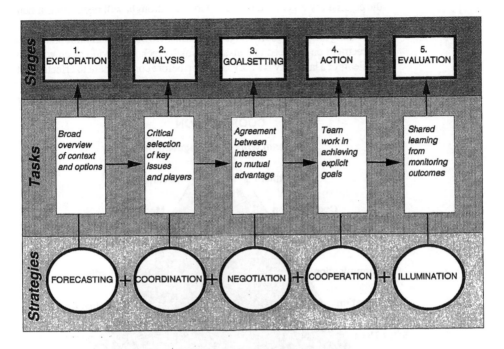

Figure 4.1 Strategies for environmental management

In Figure 4.2 conflict management is represented as a mountain to be climbed. Each of the steps up the mountain is effective, depending on where the player is in the process. Achieving a cooperative team will involve all the key players in the management process. Since all the players almost certainly live and work in a specialised, competitive society, they are almost certain to start from a professionally isolated position. Reducing the individual isolation of specialised interests can hardly be achieved by force. The first incentive for the players to work together is **information** on shared interests. This may lead on to **communication**, which will mean learning more of each others' positions and interests. Working together on the interests in common is **collaboration** when the information is shared and communication is established.

Collaboration implies a formal and possibly temporary working relationship. **Coordination** goes further in that it sets up systems of communication with agreed ground-rules. A formal meeting on water quotas during a drought may bring collaboration; monitoring the water use will need coordination. Holding the community together through a flood or a drought will need commitment to common goals.

Sharing tasks and goals brings commitment to both the outcome and the team. **Agreement** on the need to conserve a small creek or provide clean water to all residents of a large city is a necessary step before commitment to effective action can be expected. Once such cooperation exists, active **team-building** can follow, with members sharing skills and resources to their mutual advantage.

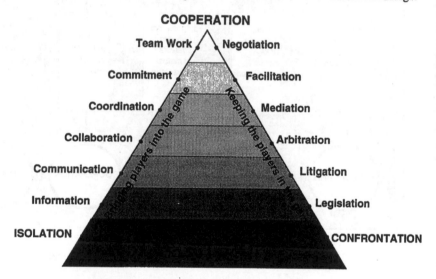

Figure 4.2   Conflict management: scaling the mountain

While all conflict management strategies are appropriate in particular circumstances, at certain stages some are more appropriate than others. A brainstorming session would unite the diverse group of interests involved in managing the example of Calico Creek. Team-building strategies would allow for recognition of mutual interests between the competing forces in Nyah Shire. In an urban area more formal committee procedures may be needed to harness the energies of the different arms of government. Cooperative team work, and mutual goalsetting, monitoring and evaluation of the work programme are all now mainstream practices for modern environmental managers.

Once a team has been formed, it will benefit greatly from joint training in negotiation and facilitation. Training in any of the conflict resolution methods can be remedial, after conflicts have arisen. Or, preferably, conflict management skills may become a standard part of routine professional training. It is more beneficial for members of the management team to develop conflict management skills for their own use than to rely on external help.

When cooperation is threatened with collapse, or if it has yet to be achieved, remedial measures will be needed to keep the players in the game. On the confrontation slope of the conflict management mountain, the task is either to reduce existing conflict (starting at the bottom) or to maintain the hard-won cooperation (stopping a slide from the top). In each case the aim is to keep the players involved in a constructive, cooperative type of game, rather than leave them free to turn to opposition or sabotage. Short of outright war, **confrontation** is the next most destructive way of approaching conflict. This is often solved by **legislation** with sanctions for disobedience. This solution applies only after those wielding power, or where the bulk of the population, have become sufficiently concerned. Regulations on the transport of intractable wastes are currently being enacted in Australia. The problem existed for decades, but was not dealt with until communities on the transport routes felt at risk.

Litigation, for example the citizen's use of the law courts, contributes to the development of environmental case law. One hundred citizens of the Australian national capital took the Federal Government to court to block a telecommunications tower which they considered to be unsightly. They won the court case, lost the battle (the tower was built) and established the first citizen's action environmental case-law in Australia. Use of the courts has become a common avenue for environmental conflict resolution. It is expensive for both society and litigants. **Arbitration**, through appealing to a specialised tribunal, reduces the cost and increases the level of expertise available for judgment. However, it requires that established agencies be available and brings with it the risk that the agency may develop a fixed agenda of its own. This has been a criticism of Australian State Land Courts. The creation of the Environmental Protection Agencies in most countries could be regarded as a step towards an arbitration court for environmental issues.

Mediation, with the use of a trained negotiator, is increasingly being applied to environmental disputes. The most well-known is the mediation processes in the water and forest industries, where they have become highly sophisticated. Mediation is a useful tool for solving entrenched problems. However, if it is introduced from outside the management process, the longer-term effect may be lost through lack of real commitment to the solution by the participants. Mediation is becoming incorporated into the new environmental legal procedures. There is a risk that it may become tainted with the adversary role central to litigation and the law courts.

The use of a facilitator is a less formal process than employing a trained mediator. **Facilitation** involves setting up and enabling a constructive, problem-solving environment for the stakeholders in an issue. The facilitator, who is usually someone perceived as neutral by all the participants, sets and keeps the groundrules until resolution is achieved. Politicians are often skilled facilitators on national issues – although they are not often given the credit. The chair of a typical committee meeting acts in a facilitator's role. They supervise standard committee procedures designed to allow every member a voice. The 'round tables' which have become an integral part of environmental management in Canada and Norway are a facilitating device designed to move participants from collaboration to teamwork.

Conflict management methods become increasingly less intrusive in the management process as one proceeds up the incline from legislation to facilitation in Figure 4.2. **Negotiation**, whether applied within the everyday processes of management, or specially convened to resolve a major issue, is a crucial strategy. Negotiation takes place between the players themselves, and is traditionally aimed toward a win-lose or win-win outcome between two sides. A more recent approach to negotiation, exemplified by the Harvard Negotiating Process, is to negotiate to the mutual advantage of all players. There is thus no need for victory or defeat. Ideally, everybody gains at no-one's expense. This is not an empty ideal but the basis for modern business management. The lead-up process to the second global conference on environment and development in Brazil in June 1992 was equivalent to a long series of negotiations. The process included five preparatory conferences between all 77 participating nations, a global conference of non-government organisations and a global conference for women.

There is much to be gained from considering the five stages of environmental management as a problem-solving process, with coordinated and constructive management of conflict as the task at each stage. The different stages of environmental decision-making are often managed separately, as impact assessment, action and evaluation respectively. Such discontinuous management is a sure source of entrenched disputes, as is shown by the number of discarded Impact Assessment Statements. An important step in reducing this potential for confrontation is to reorganise the management process into a connected series of events.

Each issue is then managed broadly, with recognition of the various components (economic, ecological, social and technical) and the full range of players (experts, stakeholders and management team). The players work cooperatively with continuing responsibilities for the decision-making on any one issue. This continuity holds from the original problem identification to the next steps after its resolution. The five main stages of environmental decision-making in Figure 4.1 hold a range of tasks, from examining the broad context of, and options for, the issue, to joint understanding of the outcomes. For each of those tasks there is an appropriate management strategy which maximises opportunities for cooperation and minimises confrontation and isolation.

A Stakeholder Analysis

The first stage of decision-making calls for a forecast of the issues by all possible stakeholders. Analysis of the result allows key players and resources to be identified, but these will need formal coordinating processes to ensure that they continue to contribute constructively to management. In Calico Creek the players may be debarred from meeting by distance. In Bangkok, the barriers will be mainly the professional and departmental territories in the bureaucracy; and the gap between rich and poor. In Nyah Shire the four partners in the administration – Federal, State, and Local Government and community – will need active intervention to achieve coordination.

In putting together a multi-skilled team representing all the interests, there are three challenges built-in before the start:

1 breaking down the prevailing norms of fragmentation and speciali-sation in disciplines and professions;

2 identifying cumulative effects whose combined consequence can only actually be observed if a threshold is crossed;

3 putting theory into practice, and thereby breaking the inaction barrier which is likely to have been erected.

The first of three challenges to managing the currently changing areas and unpredictable environmental issues is to coordinate disparate information, ie to link together the answers to the many types of questions which need to be asked. While the potential number of specialised fields is almost limitless, the particular set of questions to be asked is controlled by the framework in which any change occurs. The list of potential information sources and skills of investigation on ecological systems on the one hand, and human systems on the other, need to be coordinated within a strategic framework such as Figure 4.1.

The second challenge is met by harnessing the various interested parties in the project for collaborative, rather than conflicting ends. Various interest groups provide information on, and are capable of influencing the outcome of, the project being assessed. If they are groups closely concerned with the issues for the project, then they must also be part of their resolution. While the number of such interests will be wide, their categories are predictable. There will be key players from policymakers, experts on facets of the physical and social environment, community agencies, influential individuals, and the professions and trades which service the area (Table 4.1). A stakeholder analysis as outlined in Table 4.2 can be used to identify the main players of the significant interest groups (Brown 1994). This exercise is a useful preparation for round table consultation at all of the five stages of environmental problem solving.

Table 4.1   Knowledge and skills in environmental management

Dimension	Knowledge and skills	Aspects of environmental management
Policy	Ecology, economics, law, social security, transport, health, finance	The questions
Environment	Health markers and goals, environmental indicators	The database
Community	Interest groups, social systems, rules and goals	The context
Individuals	Behaviours, norms, values, choices	The people
Services	Finance, law health, housing, planning environment	The workforce

The third challenge is to overcome inertia and put theory into practice. A useful approach is to turn the activity from management into problem-solving. Figure 4.1 can be regarded as describing the various stages of problem identification, taking action and re-evaluation of the state of the environment. Problem solving is, after all, not a discrete tool like mapping or a survey, but is itself part of the political and planning processes. The goal setting and review processes are not necessarily in conflict with the development process (as is often assumed), but should augment it. The information collected should be at the disposal of the key interest groups while decisions are still being made. The pool of information is then available for both 'macro' overview and 'micro' special interest perspectives on the progress of the project, and accessible for the common benefit of management and community.

The Coordination Process

The first hurdle to overcome in meeting the three challenges is the confusion in the environmental literature between environmental problems and social problems. The first, environmental problems, refers to a given body of knowledge about the natural environment, with its own frameworks and experts who can be called upon to give reliable advice within their own frame of reference. The second, social problems, refers to the mixture of political intention, social expectations and aesthetic choices which inevitably accompany environmental decision-making. These decisions may draw on one field more than another, such as ecology, physiology or finance, but information may be needed from the full range of fields.

'Is this water catchment reliable?' is a question for ecologists. 'Is this water drinkable?' is a combined biological, economic and political decision. 'Is this chemical factory safe for a small town?' is a question with two parts. First, 'Does it meet established environmental standards?' and second, 'Is it economically necessary, politically feasible and acceptable to the local residents to build the factory?'. Whichever way the scientific evidence answers the first, the real life decision will await answers to the second.

For the first type of question, ecology, geology and epidemiology are principle sources of information. For the second type of question, economics, social studies and law are also crucial. In the actual integration the experts may find they have assumptions which conflict, eg the legal emphasis in health questions is on the rights of the individual, and on social justice matters; in environment it is on regulation and the protection of public rights.

For Type 2 environmental questions, some sort of coordination between the disparate information sources is essential. Table 4.3 uses the integrative framework of the Ottawa Charter, with its five action strategies (Brown 1990), as a basis for coordination. This allows identification of where and how each type of information will be used, and so pre-empts some of the traditional confusion outlined above. Use of such a table draws attention to the need to deal explicitly with policy and political questions; and that in current environmental issues we are almost certainly dealing with change.

The issue remains: how is a society such as ours, mostly trained to give, or to respect, expert advice, to supply skills in collating and coordinating evidence in this way? It is not necessary, and is even absurd, to aspire to be a specialist in every field related to a project (the 'exhausted expert' model); or to try to interpret a little of every set of information (the 'superficial generalist' model). The challenge of coordination is:

1   having decided on an appropriate coordinating framework; and
2   to select the material for its relevance to the issue, while assessing its validity in its own field, and ensuring its translation into language understandable across specialities (an 'effective editor' model).

This process takes account of the supply side of coordination of information, the potential providers of advice. It does not, however, take account of the difficulty of melding the diverse interests of parties who seek to use the advice in evaluation and management phases of the impact assessment; the demand side, as it were. A collaborative system of consultation is needed.

Table 4.2   Stakeholders in an environment issue

Dimension	Stakeholders
Policy	Prime minister, state premier, local council, government and now government agencies
Environment	Pollution control, urban planners, health surveyors, recreation officers, inhabitants
Community	Farmers, parents, residents, unemployed, church, industry management
Individuals	Membership, age groups, socio-economic levels, ethnic groups, home owners, underclasses
Services	Soil conservation, urban planning, teachers, police, lawyers

Table 4.2 offers the basis for an analysis which can be used to develop collaboration and/or at the evaluative stage, where the first management questions arise. It assumes that every interest group in the community affected by the project is a potential stakeholder; that only about twelve representatives in practice become deeply involved; and more usually only between five and seven. These twelve will always include:

- financial and political backers of the project;
- experts on technical matters such as water quality, dam safety etc;
- representatives of regulatory authorities;
- consumers of the service offered by the project, usually industry;
- representatives of government (local, State and/or Federal); and
- local community leaders.

Coordinating Power and Resources

The resources available to each interest group include at least five potential sources of power and influence for each one, namely, the social authority; their particular knowledge base; their political power through their alliances; and personal power in having been chosen as a representative; and their particular working style.

Identifying the potential resources for each stakeholder in the project acts as a project resource in itself. It demystifies the power plays which will inevitably be exerted during decision-making. It allows each representative to identify their own resources, and provides a collaborative basis for round table discussion and problem solving. The challenge of synthesising a wide range of collated information has been variously defined as a knowledge issue, a social issue, a matter of language or a question of reliable evidence.

It is clearly all of these things, and a question of who is wielding which type of power. A coordinated response needs some formal system which promotes collaboration between the relevant interest groups, and builds bridges between their knowledge bases, their social context, communicating languages, and the relative emphasis they give the various faces of 'the truth'.

After the interest groups have been determined, selection of common goals encourages a mutual interest in the same creek, shire or city. Negotiation techniques with positive outcomes for all concerned are essential to bypass fixed positions and personal rivalries and invent fresh options in which all can gain. Once Bangkok can arrive at a shared set of goals for the city's future transport needs, the technical means to design and build it is a comparatively straightforward procedure. The resources needed are, on the other hand, a political matter. But if everyone has agreed, there is likely to be a political will.

In cooperative committee meetings, for instance, the options for Nyah Shire could be broadened from the current proposals of laser levelling (which is a technological fix only) or closing down agriculture (which is defeatist). Canvassing all the interested parties would widen the options considerably. The smaller management group would then negotiate in selecting the most appropriate option and provide a coordinated management plan for wider discussion within the community. The mode of negotiation applied in the Harvard Negotiating Process separates the personalities from the problem; and the fixed positions of the players from their underlying interests (Fisher and Ury 1987). This should already have been achieved in the first two stages – **exploration** and **analysis** – through sharing future visions and identifying common goals.

Negotiations within the management team continue, based on the common interests, not fixed positions, and allow fresh options to be considered. In Nyah Shire this could include a bank-supported buy-out plus voluntary rotation of paddocks and tree-planting by those who chose to remain. In choosing this option, objective monitoring techniques would be crucial to its success, as would be mutual long-term cooperation between residents and the four levels of management. Monitoring would be constant and mutual, so that some of the elements of a 'commons' would be introduced. Environmental management becomes management of a shared resource.

Table 4.3 Environmental management = conflict management

Stages	Resources
1 Exploring the issues:	by forecasting the future
2 Analysing the key factors:	by coordinating key players
3 Goalsetting for top priorities:	by negotiating to win/win
4 Acting to reach the agreed goal:	by cooperating with colleagues
5 Evaluating the outcomes:	by illuminating both intended and unintended results

While Table 4.2 and Figure 4.1 identify the conflict management resources most appropriate for each stage of environmental management, this does not exclude the use of any strategy at any stage. It is rather a case of identifying the logical resources for each main task (Table 4.3).

Using this approach it might be possible for Calico Creek to develop its resources and maintain its current quality of life. Nyah Shire may find that the pain of redevelopment can be dramatically reduced through a cooperative, rather than a punitive solution. Bangkok may develop a public management system which has goals of social equity, economic efficiency as well as of sustainable development. These rather idealised outcomes are at the least more likely with good conflict management.

References and Further Reading

Atkins, R A (1984) 'Comparative analysis of the utility of EIA methods', in Clark et al (eds) *Perspectives on environmental impact assessment* D Reidel Publishing Company, Dordrecht

Axelrod, R (1984) *The evolution of cooperation* Basic Books, New York

Brown, V (1985) Towards an epidemiology of health A basis for planning community health programs *Health Policy*

Brown, V (1994) 'Integrating health and environment: A. A common framework, and B. A common practice', in Chu, Coria Chu and Simpson, Rod (eds) *An Ecological Public Health*, Toronto

Day, D (1991) 'Resolving conflict in the New South Wales water sector: informality, anomaly and innovation', in Handmer, J W, Dorcey, A J, and Smith, D I (1991), *Negotiating water: resolving conflict in Australian water management Cres Paper 3* Centre for Resource and Environmental Studies, Australian National University, Canberra

Dorcey, A H J (1991) 'Conflict resolution in natural resources management: sustainable development and negotiation', in Handmer, J W, Dorcey, A J, and Smith, D I (1991), *Negotiating water: resolving conflict in Australian water management* Centre for Resource and Environmental Studies, Australian National University, Canberra, pp 20–46

Fisher, R and Ury, W (1987) *Getting to yes: negotiating agreement without giving in* Arrow, New York

Handmer, J W, Dorcey, A J, and Smith, D I (1991) *Negotiating water: resolving conflict in Australian water management* Centre for Resource and Environmental Studies, Australian National University, Canberra

Jarman, A (1991) 'Conflict definition and resolution: a contingency theory perspective', in Handmer et al *Negotiating water*, pp 115–124

Ury, G *(*1991) *Getting past no: negotiating with difficult people* Business Books, London

Chapters 10–16 of this volume

5   Who Manages?

On becoming an environmental manager

When environmental management equals constructive conflict management, the question 'Who is the manager' is answered by 'Whoever is involved in the issue itself'. That is not to suggest that anarchy must reign, or even that the process is always democratic. But it does confirm that where coordinated, cooperative processes are in place everybody in the affected community is contributing to environmental management. Reliance on a hierarchical or broken chain of responsibility is linked to a hierarchical and fragmented view of the environment. In the current highly specialised world responsibility is sharply divided, both vertically (between different levels of management), and horizontally (across the range of professions and interests). A partial sum of the fragmentation is indicated in Figure 5.1. Thus the constructive conflict management strategies and skills listed in Figure 5.1 lie outside of much of current management experience.

It would be irresponsible to move to integrated environmental decision-making without acknowledging the considerable individual and organisational learning necessary for it to run smoothly. As well as being conflict management, modern environmental management is also management of personal and social change. This cannot be pre-ordained for any one person, nor any one organisation. Every adult already has a personal style, a rich range of experience and previous training. Each person contributing to environmental management, which includes the entire range of participants indicated in Figure 5.1, will need to build on their existing capacities, not dismiss them. Organisations have spent years or decades constructing their institutional styles and this will not be lightly redirected.

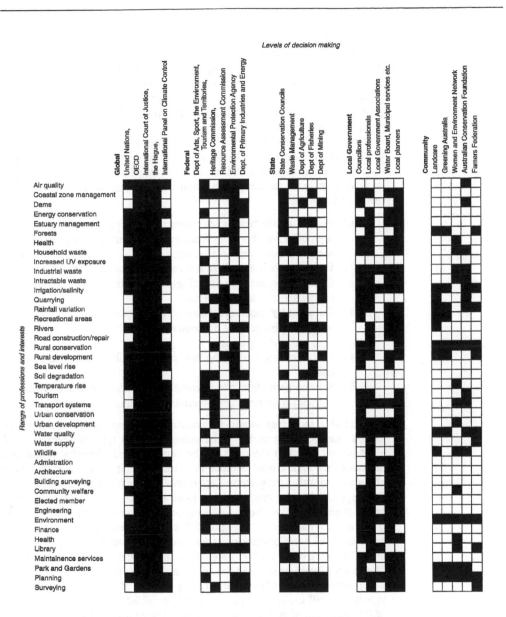

Figure 5.1 Elements of environmental decision making

The new environmental managers must develop their own personal learning styles and acquire their own management skills and techniques. The traditional in-service education model would update skills in the specialist groups in which people were originally trained, or else in extension courses on particular environmental issues. Figure 5.1 offers evidence of the number of different programmes needed on this specialist education model. The fragmentation which everyone agrees is part of the problem would then become embedded in the solution. It would be of far greater value to group people on in-service courses in the spheres in which they manage. Five such spheres incorporate the principal dimensions of management and all the elements of society needed to achieve social change. In the field of public health, the five components of the management of change are:

1 **Policy makers** who develop integrated environmental policies, linking policy makers in primary and secondary industry, community services, education, finance, etc;

2 **Research workers** who enhance the understanding of environments through monitoring their ecological, economic, demographic, social and technological dimensions;

3 **Community members** who strengthen community action by representing the interests of different regions, political positions, genders, ages and ethnic groups;

4 **Individuals** who develop their personal skills in managing change; and

5 **Service providers** and **administrators** who redirect services and agencies from dealing with effects to dealing with causes.

Considerable resources would be needed to develop specialised environmental management training programmes for each type of contributor to management listed in Figure 5.1. The very size and diversity of the grid indicates that such a goal is close to absurdity. It is far simpler and more effective to develop programmes appropriate to the players from each of the five aspects of social change. Even more effective is a programme which brings together all five. Such a coordinated programme broadens the experience of each participant. In Figure 5.2, a central theme of the five stages of environmental decision-making is just such coordination. At the first stage, all the issues are put on the table. Next, all the players come to the table. Then all the interests share in the solution. All the players join in the action which achieves the solution, and finally, all the players review the game together. Small groups of nine or ten players, with membership from each of the five types of actor, replicate the conditions needed for change.

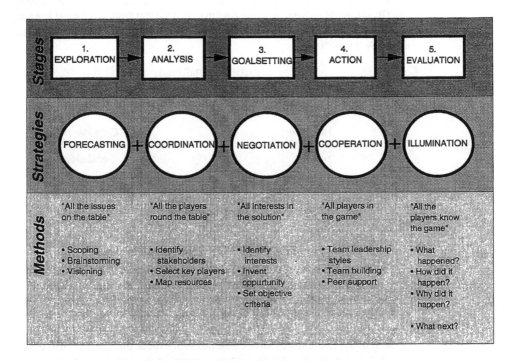

Figure 5.2 Environmental management: skills and techniques

The learning modules which make up Part 3 of this volume are designed on adult learning principles. The eight modules of the programme offer the opportunity to apply the strategies and skills listed in Figures 5.2, 5.3 and Table 5.1. Learning groups could be set up within an existing organisation, around a common issue, or across types of institutions. The learning modules could be used as the project work in undergraduate or graduate classes, or could provide the basis for design of an integrated research project. In each case, running the programme would provide further educational opportunities for those involved, as well as advancing organisational change.

Some general principles apply to all those involved in changing environmental management, whatever their role or training. These are illustrated in Figure 5.3, the Adult Learning Cycle. It is a matter of importance to all learners to identify the conditions which maximise their own capacity to learn. They may need to identify once useful skills that are no longer relevant. The classic principles of adult learning, so different from childhood learning, are a matter of commonsense logic which is confirmed by considerable educational research.

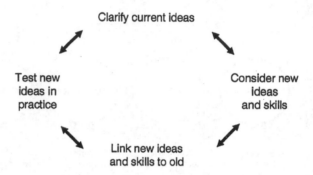

Figure 5.3 The cycle of adult learning

The Adult Learning Cycle supplies the basic principles for the learning programme which follows. At the start it is assumed that every adult is already doing their best in their own particular circumstances. Learning is change and the changes have to start wherever each person happens to be. A review of the current situation is a first step. The nature of the desired change is a matter for each individual. Only by an accurate and sympathetic diagnosis of their present position is it possible for the learner to accept or even recognise new material. If material is offered in a way which does not fit their current framework, the new material is literally incomprehensible. It is of little use to talk about interpreting overseas market trends to a farmer whose whole experience is with local self-sufficiency. Having reviewed their current position, learners can then welcome new material: the opportunity for change is great.

The presentation of material is crucial to learning. By the time they become adults, each person has a firmly acquired perceptual style, management style and learning style. This is not to say that for any individual these styles are fixed. Adding to one's repertoire of styles may be the purpose of the learning contract that learners make with themselves. The knowledge of their own and others' learning styles can in itself be valuable to a learner. An engineer consulting with the community on potential flood damage will have a specialist interest in the topic and so is likely to be convergent in management style, and favour auditory channels of communication. The members of the community on the other hand, will have a generalist interest in all aspects of the issue. They are likely to wish that the situation be managed adaptively. Any group of people will contain some who use visual and some who use auditory channels of communication.

Three forms of personal style are particularly relevant in changing environmental management. Whatever the individuals' training or background, whether engineering, administration or health, they will have approached information principally as specialists, generalists or holists.

1 Some learners may have a preference for collection of information in small accurate pieces, within a given framework with which they are thoroughly familiar. This is the **specialist's approach**.
2 Others may wish to achieve a **generalist's understanding** of the whole of a topic or issue. They will be prepared to range more widely in collecting information needed for specialised review.
3 This group also wishes to understand the whole of an issue, but finds that they cannot do so until they have developed a framework which will explain the information they are gathering (the **holistic approach**).

Almost all contemporary education is in the specialist mode; whereas many of the information needs of environmental management are better met by using generalist and holistic skills.

Specialists, Generalists and Holists

'What's in a name? A rose by any other name would still smell as sweet'
Shakespeare, Romeo to Juliet

Or would it? To a botanist, a rose is one of a great number of species in the genus *Rosa*, part of the Family *Rosacae*. Related to the apple, it is characterised by compound leaves, a five-fold petal arrangement, multiple stamens and inferior ovaries. By contrast, to a suburban gardener it may be the reward for living on a quarter-acre block and strongly resisting urban consolidation. A florist sees a rose as one element in a living picture. Its colour, scent and shape need to complement the flowers around it. In many cultures, roses carry meaning beyond mere biology – such as red roses for love and white for purity – and so act as important social symbols which everyone can recognise at a glance. Red and white roses are still historical symbols of opposing sides in a war – the York and Lancaster battles for the English throne in the Middle Ages.

Quite different perceptions of an object as familiar as a rose reflect the three broad strategies people have for coping with ideas, information and facts – as specialists, generalists and holists. They are approaches we all use at different times. However, everyone has one method they prefer to use, or their training has developed. In current western culture it is most likely to be that of specialist.

Broadly speaking, a specialist is someone interested in specific knowledge in one area of a particular subject. The botanical description of a rose concentrates on the biological and taxonomic characteristics of roses – it is unconcerned with whatever emotions humans attach to them. By contrast, a generalist is interested in all forms of information that can be obtained about a subject, regardless of how different pieces of information are related to each other. A holist is also interested in all kinds of information, but needs to arrange it into a cohesive structure. The giver of flowers is concerned with the message that the rose transmits, and that it is clearly understood by the receiver.

These characteristics how people tend to behave are generalisations, of course. No one action fits perfectly or exclusively into any one of these characterisations. We all use these different approaches at different times, each where it is perceived as appropriate. To broaden the example beyond botanists, gardeners and florists, consider the different approaches that each type of person would take to obtain information about a particular topic.

Table 5.1   Methods of inquiry: specialist, generalist and holist

	Specialist	**Generalist**	**Holist**
Type of question	What aspect am I qualified to examine?	What do the various specialists have to say?	How can I best understand the whole?
Subject matter	Specialist topics– eg species, artefacts, behaviours	Summaries or interpretations and new questions	First order questions
Method of inquiry	Specific to the specialism – measurement observation or interpretation	Reading, discussion and criticism	Eclectic – qualitative and quantitative
Source of 'truth'	Results of accepted method of inquiry	Accepts specialist opinion	Eclectic – each type of evidence validated separately

What, then, of the way that people are thought to handle information, and the way that it is presented to us?

In western culture, for instance, our universities – the pinnacle of our formal learning system – are traditionally divided into the sciences and the liberal arts. Over the past three centuries, the sciences have become increasingly – some would say overly – specialised. The study of the transfer of potassium ions through the cell membranes of *E coli* may seem like a ridiculously minute area of curiosity to some, but for a cell membrane scientist, it may be an important area of study, with significant ramifications. Scientists and associated professionals are specialists *par excellence*.

The power of science has come from the enormously detailed understanding that it has accumulated on every conceivable subject: the result of hundreds of years of study into the physical nature of the universe. The social sciences by contrast have been more concerned with the diversity of the human experience. For instance, politics is not a subject that lends itself to systematic dissection, but rather a broad survey. An understanding of political theory is bound up with an understanding of history, economics and social conditions. The evolution of Marxist theory, for instance, cannot be understood by dissecting *Das Kapital*: understanding Marx involves understanding the increasing industrialisation of 19th century Germany, the autocratic nature of the Second Reich and the complex relationship of Marx and Engels.

For many, the arts and the sciences are the epitome of the schools of generalist thinking and specialist thinking respectively. Until recently, there was little traffic between the two, which has made the third type of coordinating information, the holist seeking synthesis, the most difficult to achieve.

A specific example of the deep schism that has grown up between the two is the division of library facilities. Normally they are separated, on different floors, if not in different buildings. The designers of these facilities obviously expected there to be little of common interest between the two – and unconsciously ensured that it would be so by separating them. With the rise of computer-based access systems, the physical reasons for the division are gone, but the academic division remains.

A study conducted at an Australian university provides some evidence on specialists, generalists and holists, and the way they dealt with information. Seven hundred and fifty students were asked about their thinking and learning, and their practical work and problem-solving was observed. Some of the results of this survey are shown below. The results show a clear difference in the way that each group processes information (Brown 1978).

Perhaps one of the most important results from this survey is the people that each group had met. While most had come hoping to meet a wide range of people, only the holists found that they had (not shown in Table 5.2). The

65

generalists and specialists found themselves surrounded with people that think the same way as themselves. Have we not only three information processing styles, but three personality types? The research with this group of students seems to advance this conjecture.

Table 5.2 Specialist, generalist and holist approaches to knowledge

	Specialist (N=345)	Generalist (N=152)	Holist (N=244)
Main reason for entering university	*professional training (%)*	*general education (%)*	*subject interest (%)*
Meet a wide range of people	71	73	77
Knowledge relevant to today	43	55	71
Use of research journals	15	20	56
Wish a wide range of information	70	68	87
Pragmatism (z-scores)	0.02	0.36	0.28
Tolerance of complexity	0.06	-0.03	0.19
Intolerance of ambiguity	0.15	0.02	0.08

Of the three groups, only the holists were interested in the relevance of the knowledge that they are being given, expected to use research journals and recent findings extensively, covered a wide range of information, and were tolerant of complexity in solving problems and dealing with social issues. By contrast, specialists expected to obtain the majority of information from established texts (not research papers), were less concerned whether their knowledge was useful or not, did not readily tolerate complexity, and were more likely to be intolerant of uncertainty and ambiguity.

The problems of universities – in information and people – is a microcosm of problems in the wider community. More importantly, universities are perpetuating the problems. For the western world at least, the vast majority of our leaders, managers and high level decision-makers will pass through the university system. They will be inculcated into the pervasive paradigm: division. This is not a problem in itself, so long as the contribution of this division to the perpetuation of conflict is recognised. Resolution of conflict is a holist activity – drawing together all parts within a single explanatory system.

Those who seek explanations from a wide range of specialists are often excluded from academic debate. They are excluded by the same means that lead to the positions of only one view (only Europeans', or only males', or only engineers') being heard in an environmental management strategy.

In the context of environmental management, the case becomes not only a matter of principle, but deep practical concern. The environment functions as a hugely complicated organism made of many equally complicated units. To understand the environment, it is essential to understand how each piece works – a specialist's skill. It is also important to identify all the units within the organism – the place of the generalist. And finally it is imperative to understand how the parts interact with one another – the role of the holist. It is clear that managing the environment cannot possibly be successful without all three; and the third is the role of the conflict manager.

And to answer Shakespeare's question at the outset of this paper – *What's in a name? A rose by any other name would still smell as sweet.* The rose does not care what name people might call it – but to humans the name is critical. It is the rose known by *all* its names that is the basic step towards conflict management and resolution.

Adapting, Accommodating, Converging and Diverging

Management style refers to the ways in which individuals solve problems and make decisions. A management style often but not always matches a problem-solving style. One useful analysis of management styles identifies four orientations: ADAPTATION, ACCOMMODATION, CONVERGENCE and DIVERGENCE (Kolb 1974). The four styles correlate well with the occupations people choose. For instance, there are many engineers with convergent management styles, and administrators with accommodating (adaptable) styles. In an environmental management team, it is important to have both. It is also important to know that their styles are different, but complementary.

It is unlikely that anyone's problem-solving style will be described entirely by any one of the four management styles alone. This is because each person's problem-solving style is a combination of the four basic learning steps approaches.

- **The converger**: the convergers' dominant problem-solving abilities are to construct abstract concepts and actively experiment. Their greatest strength lies in the practical application of ideas. People with this style seem to do best in those situations like conventional intelligence tests where there is a single correct answer or solution to a question or problem. Their knowledge is organised in such a way that they can focus it on specific problems. Research on this style of learning shows that convergers are relatively unemotional, preferring to deal with things rather than people. They tend to have technical interests, and choose to specialise in the physical sciences. This learning style is characteristic of many engineers.
- **Divergers** have the opposite learning strengths of the convergers. These people are best at concrete experience and reflective observation. They will excel in the ability to view concrete situations from many perspectives. They perform better in situations that call for the generation of ideas, such as 'brainstorming' sessions. Research shows that divergers are interested in people and tend to be imaginative and emotional. They have broad cultural interests and tend to specialise in the arts. This style is characteristic of individuals from humanities and liberal arts backgrounds. Counsellors, organisation development specialists and personnel managers tend to be characterised by this problem-solving style.
- **The assimilators'** dominant problem-solving abilities are a combination of constructing abstract concepts and making reflective observations. Their greatest strength lies in the ability to create theoretical models. These people excel in inductive reasoning and in assimilating disparate observations into an integrated explanation. They, like the converger, are less interested in people and more concerned with abstract concepts, but are less concerned with the practical use of theories. For them it is more important that the theory be logically sound and precise; in a situation where a theory or plan does not fit the 'facts', the assimilator would be likely to disregard or reexamine the facts. This problem-solving style is found most often in the research and planning departments.

- **Accommodators** have opposite learning strengths to the assimilator. These people are best at concrete experience and active experimentation. Their greatest strength lies in doing things – in carrying out plans and experiments – and they like being involved in new experiences. They tend to take more risks than people in the other three problem-solving categories. They are labelled 'accommodators' because they tend to excel in those situations where one must adapt oneself to specific immediate circumstances. In situations where a theory or plan does not fit the 'facts', they will most likely discard the plan or theory. They tend to solve problems in an intuitive trial-and-error manner, relying heavily on other people for information rather than on their own analytical ability. The accommodator is at ease with people but is sometimes seen as impatient and 'pushy'. Their educational background is often in technical or practical fields such as business. In organisations people with this learning style are found in 'action-oriented' jobs, often in marketing or new industries.

Visual, Auditory, Experiential

Individuals are highly selective in the way they receive information. Of the four main channels through which we receive information – eyes, ears, smell and touch – individuals have a preferred mode for receiving and for sending (Bander and Grindler 1981). Smell is an underused mode in modern urban life. But structured exercises quickly demonstrate how biased are individuals' preferences for receiving information through sight, sound or feelings. While the engineer and the financial manager are likely to prefer auditory channels, the planner and the health-care provider are likely to prefer the visual. The community worker and the environmental activist are likely to perceive through their feelings.

These analyses have several advantages in the management of change. Both oneself and one's work colleagues will be accessing new information through all three learning strategies. It can be important to take account of the different strategies which are being used, and not to assume that any one person is the same as another; or worse, the same as oneself. In managing conflict, extending the knowledge of the different strategies can be used to broaden the players' understanding of the issue and their acceptance of each other's position.

Table 5.3 Adult learning styles

Information processing style	Specialist
	Holist
	Generalist
Management style	Adaptation
	Accommodation
	Convergence
	Divergence
Information reception channel	Visual
	Auditory
	Olfactory
	Experiential

Organisational change is addressed in Chapter 16, and deals with the re-entry problem. For an individual to apply newly acquired ideas and skills in their workplace, the workplace itself has to change, or at least be receptive to change. Included in this volume as resources for participants in the programme are accounts of the programme as applied in a wide range of circumstances. These include an academic unit for military training, a community action programme by a shire engineer, a planning procedure for a city transport authority, and a public education programme for a waste management authority. The programme has also been applied in Nyah Shire and Calico Creek, the subjects of two of the case studies in Part 2, and in a water catchment managed by the traditional Aboriginal owners.

To complete their learning cycle, adult learners will need to apply their learning in their usual working context. Only then can they determine whether the new skills have become part of their own repertoire, or were merely part of a short-term diversion. In this sourcebook, each of the seven modules offers the opportunity to test the skills listed in Table 5.4. The first five modules cover the five stages of environmental decision-making. After the sixth session, in which the supports and barriers which meet the learner back in the workplace are identified, the seventh session allows each participant to review their own learning and decide the best way to develop it further.

Table 5.4 Environmental conflict management: personal skills

	Resources	Skills
1	Forecasting	Scoping Brainstorming Visioning
2	Coordinating	Identifying stakeholders Selecting key players Mapping resources
3	Negotiating	Focus on the interests Invent options Set objective criteria
4	Cooperating	Team leadership styles Team building Peer support
5	Illuminating	What happened? How did it happen? Why did it happen? What next?

In changing environmental management the most valuable of all resources are the experiences of each team member. This book, with its learning modules and case studies of water resource management, is intended to provide a microcosm of environmental management as it is today. The actual learning takes place in one place only in the mind of the learner, and the interrelated behaviour of the management team.

References and Further Reading

Part 3 of this volume.

Brown, V A (1978) *Holism in the university curriculum: promise or performance?* Australian National University, Canberra

Grinder, J, Bandler, R (1981) *The structure of magic 1. Neurolinguistic programming* Science and Behaviour Books, New York

Kolb, D A, Rubin, I M and McIntyre, J M (1974) *Organizational psychology: a book of readings* Prentice-Hall Inc, Englewood Cliffs

Kuhn, T (1970) *The structure of scientific revolutions* Routledge, London

Maturana, H (1988) *The tree of knowledge: the biological roots of human understanding* Shambhala, Boston

Part 2

Managing Environmental Change

6 A Heroic Effort

A case study of the salmon fishing industry in British Columbia, Canada*

D Ingle Smith and John Handmer

Background

Canada is a nation that abounds with natural resources. The overlay of modern society, with its insatiable appetite for natural resource products, is still at a stage where it appears that management of the environment could be undertaken within the bounds of sustainable development. Nevertheless, viewed over the long term, salmon fishing in the Fraser River basin of Canada has been an environmental disaster: over the last century salmon numbers have been reduced by 80 per cent. The conflicts of the last few decades, the management mechanisms put in place and the optimism over the future concern the remnants of a

---

* Canada has been at the forefront of nations responding to the challenge of *Our Common Future*, the report of the World Commission on Environment and Development – the Bruntland Commission (1987). The Canadian federal government approach was set by the publication of *Report of the National Task Force on Environment and Economy* (1987). In British Columbia this challenge was taken up by Westwater at the University of British Columbia. They obtained funding from federal and provincial government sources to undertake a major two-year study into the futures of the Fraser River basin. These were published in two volumes in 1991, entitled *Perspectives on Sustainable Development in Water Management: Towards Agreement in the Fraser River Basin* (ed Dorcey 1991a) and *Water in Sustainable Development: Exploring our Common Future in the Fraser River Basin* (ed Dorcey 1991b).

once abundant resource. The fact that the Fraser is arguably still the greatest salmon river in the world is a sad comment on the fate of salmon and their habitats elsewhere. Nevertheless, the Fraser is one of the few major salmon rivers with no dams on its main stream – a decision in favour of the fish.

The western province of British Columbia can be regarded as a microcosm of the nation. It is well-watered, rich in forests and wildlife and with a diversity of minerals. The heartland of the province is the Fraser River Basin (Figure 6.1). The main stem of the river is 1325 kms long with a total catchment area of 234,000 kms (about half the size of France). Within its catchment there is a diversity of climate and vegetation. The inland population is sparse, in contrast to the metropolis of Greater Vancouver located on the Fraser estuary and supporting 1.2 million out of the total Basin population of 2.2 million.

The Basin, including Vancouver, supports almost two-thirds of the British Columbia's population and is the source of:

Table 6.1 The Fraser River Basin: sources provided

	%
Gross provincial product	80
Sockeye salmon catch	66
Pink salmon catch	60
Metal mine production	60
Sport fishery catch of the province	49
Operable forests	48
Long-run sustainable yield of timber	46
Provincial farmland	44

In the next 25 years the population of the basin is expected to increase by 50 per cent. As with the current population distribution, the increases will be uneven and key urban areas may well double their populations in this period.

These initiatives have advanced still further by the establishment, in May 1992, of the Agreement Respecting the Fraser Basin Management Program. This is a five-year co-operative study: '... for coordinating and harmonizing government and private sector initiatives to ensure the sustainability of the Fraser Basin ... recognizing the need for unprecedented co-operation and partnership among federal, provincial, local First Nation [native peoples] and non-government stakeholders in the Basin.'

The case study reported here is based on the two Westwater volumes. We are particularly grateful to Professor A H J Dorcey for his encouragement and for permission to use the material. Citations to individual chapters which are particularly relevant to the salmon resource are listed in the references at the end of this chapter.

Settlement and governance of many of the resource rich regions of the world were superimposed by colonising European powers upon long-established native communities. This is true of British Columbia and the Fraser Basin. The First Nations of the world, the indigenous peoples, after long neglect, are now gaining a voice in the allocation and development of natural resources. They are specifically accorded status in the Bruntland Report *Our common future* (WCED 1987). This is on the grounds not only of equity but because of the lessons for sustainability that the First World can learn from the First Nations.

The disappearance of indigenous or tribal communities is a loss for the larger society, which could learn a great deal from their traditional skills in sustainably managing very complex ecological systems. It is a terrible irony that as formal development reaches more deeply into rainforests, deserts and other isolated environments, it tends to destroy the only cultures that have proved able to thrive in these environments (WCED 1987 pp114). The opportunity afforded by such lessons is only matched by the scope for conflict that the very different approaches, objectives and value systems engender.

Salmon and Early History

The consensus is that Pacific salmon and the native peoples of the Fraser River have been present in the Basin for the past 10,000 years. The salmon were an integral part of the diet and culture of the inhabitants and there was balance between fish and native communities – an example of ecologically sustainable development. The depletion of the resource, and ensuing conflict, is undoubtedly a result of European settlement. However, details of fish stocks from the early period will always remain a matter of scientific conjecture. There is more information available on changes to the environment, all of which have, to an uncertain degree, worked against the sustainability of the Pacific salmon resource base.

Data on fish populations are notoriously difficult to obtain. Unlike trees they do not stand still to be counted! However, information on the stocks, catches and movement of Pacific salmon in the Fraser Basin, from the early 1950s, are probably among the most complete statistics available anywhere. The best available estimates of historical data, for the period 1800 to the early 1900s, suggest that the numbers of salmon have experienced a five-fold decrease when compared to the period 1951 through to the early 1980s. Estimates of average annual abundance for pink and sockeye salmon are given below (Table 6.2).

Table 6.2   Estimates of annual abundance of salmon

	Historical (1800s to early 1900s)	Recent (1951–early 1980s)
Pink	23.85	4.32
Sockeye	34.23	6.75

Number of salmon in millions

The Hudson's Bay Company had established a network of trading posts through the Fraser Basin by the early 1800s. An important element in the food supply for these newcomers was dried fish traded with the native Indians. This is not thought to have adversely effected the sustainability of the resource, not least because the native population had already been reduced in numbers from introduced diseases such as measles and smallpox ahead of the arrival of the traders.

Commercial fishing for the local market increased during the century with a key issue the development of fish canneries which started in the 1860s. This extended to serve an international clientele. The commercial fisheries that fed this new industry were both marine and estuarine. By about 1900 the commercial fisherman had been joined by recreational fishers. For salmon much of the commercial and recreational fishing is at sea or near coastal waters.

Government and Salmon

The Canadian federal constitution of 1867 vested primary control and ownership of most natural resources (including water) in the Provinces – although the national government retains broad powers which may override provincial authority. The British Columbian *Fisheries Act (1856)* pre-dated federation. This early legislation was motivated by concern for overfishing and included the provision of licenses, fish replenishment, hatcheries and so on. With Federation, conflict emerged between the provincial and federal governments over the jurisdiction and control of salmon.

This early conflict was finally resolved in 1920, following a decision by the Privy Council on appeal from the Canadian Supreme Court. (This was the third time the Privy Council had dealt with federal/provincial conflicts over salmon.) The federal Department of Fisheries and Oceans (DFO) is responsible not only for inland and coastal salmon but for bi-national treaties with the USA. The most recent, in 1985, is the Pacific Salmon Treaty. This incorporates pre-season and in-season negotiations for fish allocation and management. The Fraser Panel is a component of the organisational arrangements for this purpose.

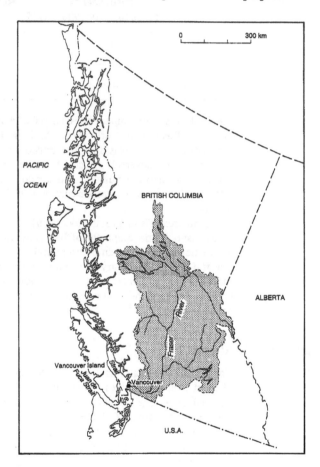

Figure 6.1   The Fraser River

From last century until well after World War Two the DFO, and its federal forerunners, was the dominant agency. Its authority is that of world leader in fish biology, its goal to increase salmon numbers within a conservation ethic. Its power lay in the ultimate decision making power of its Minister. The *Water*

Act includes sections that define offences as carrying out any work that results in '... harmful alteration, disruption or destruction' of fish habitat and disallows the deposit of '... any deleterious substance' into waters frequented by fish. These potentially strong controls have not been fully used for a range of political and practical reasons. But they do give the minister the authority to participate in the assessment of a range of developments.

This overriding dominance has been challenged on two fronts. First, that resource economics offers an alternative view of 'sustainability' and secondly, the question of the rights of indigenous peoples.

The Pearse Commission report in 1982, *Turning the Tide: a New Policy for Canada's Pacific Fisheries*, questioned the conservation role of the DFO and replaced it with the concepts of 'maximum sustainable yield' and 'limited entry'. The Commission hearings also acted as the focus to orchestrate new voices along with those of long-established interest groups such as the United Fishermen and Allied Workers Union (UFAWU).

The emergence of the role of aboriginal communities was a little later. In 1973 the Supreme Court of Canada held that aboriginal rights were part of the law of the land. However, in the context of the use of the salmon resource in the Fraser Basin definitions are still awaited as to what this means for aboriginal title and self-government. The Pacific Salmon treaty allocates 400,000 fish annually for native use, but the issues are much wider than that.

A landmark was the Sparrow decision. Sparrow, a Musqueam Indian fishing in the Fraser Estuary, was charged with the use of a drift net that exceeded the length allowed under the Fisheries Act. In 1990 the Supreme Court of Canada upheld Sparrow's aboriginal right to fish for food, social and ceremonial purposes notwithstanding the licence specifications of the Fisheries Act. Whether this could be extended to commercial fishing was not an issue before the court. The Sparrow case, however, heralded the dawn of a new era.

The DFO has recognised the need for change. Proposals in *Vision 2000 – A Vision of Pacific Fisheries at the Beginning of the 21st Century* (DFO, 1989) make this clear:

> Sustainable development can only be achieved through the coordinated efforts of all levels of government, industry and citizen groups. DFO is working to build partnerships and strengthen decision-making institutions to help promote the exchange of views and the building of a consensus for action among all key groups (DFO 1989 pp 11).

Historically, the DFO was the arbitrator of equity on behalf of the Canadian government. Its strength was its scientific expertise, its conservation ethic and the ultimate decision-making power of the Minister. It has progressively espoused

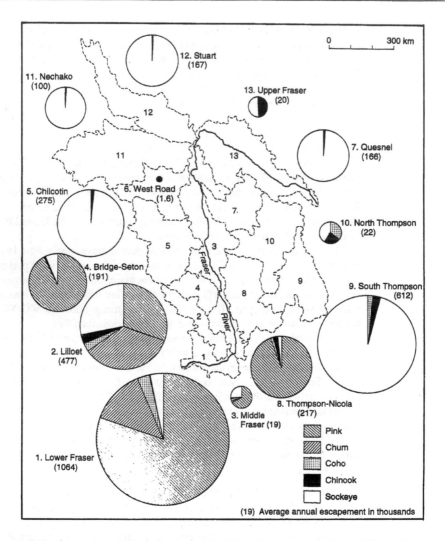

Figure 6.2 Average annual escapement of Pacific salmon to sub-basins within the Fraser River Basin (from Northcote and Burwash 1991)

policies of 'co-management' and 'consensus', although it finds problems in relinquishing its supremacy – unavoidable in truly balanced negotiations – and its conservation goals.

Environment and Salmon in the Fraser Basin

The salmon runs of the Fraser River are of universal renown. At the turn of the century the Basin was the largest world producer of salmon. The stocks are composed of six species of Pacific salmon, the popular names of which are chinook, choko, chum, steelhead, pink and sockeye. All are anadromous, that is, adults migrate upriver from ocean waters to spawn in their natal streams. Unlike their Atlantic cousins, Pacific salmon spawn only once and die once this task is complete. There are, however, significant differences between the six Fraser River species. The freshwater stage, after hatching, varies from a few days to years. This account will concentrate upon the pink (*Oncorhynchus gorbuscha*) and sockeye (*Oncorhynchus nerka*), which are numerically the dominant species. The former attain an average adult weight of about 2 kg, the smallest of the Fraser salmon species, and the latter about 3 kg.

The basic requirements of the various species of Pacific salmon are similar. The key factors are that:

- rivers have sufficient flow to allow upstream migration of the adults to the spawning grounds;
- high velocities and river blockages are absent. A total blockage such as a large dam or landslide will stop the migration altogether and result in a serious reduction in salmon numbers, possibly for years. This problem can be largely overcome by the provision of 'fish ladders', provided there is sufficient streamflow down the 'ladder'. Where blockages are partial or where high velocities exist, migration is hindered by the increase in the energy required by adult fish who do not feed in freshwater;
- gravel-bedded channels are widespread. These are essential for the redds (or 'nests'). Increased silt decreases the survival rate by blocking the interstices and thereby limiting the benthic habitat and the over-wintering refuges for the salmon fry;
- the water is of high quality. Decreases in all forms of water quality limit numbers of fish, and in extreme cases may act as barriers to migration.

However, a full understanding of the problems of managing the fish resource requires additional information. The species of Pacific salmon in the Fraser differ in several respects. These include:

- the time spent in freshwater after hatching;
- the abundance by catchment;
- the timing of the salmon runs, both upstream and downstream.

Fish census data for pink and sockeye salmon, certainly since 1951, are the equal of comparable information anywhere. Estimates of the average annual escapement (movement of adult salmon upstream) are complicated by cyclic variation in numbers. This is especially so for pink salmon which exhibit a four-year cycle and for the sockeye with an even more marked two-year variation. The annual estimates in this account have been adjusted to allow for these cyclic variations.

Figure 6.2 shows the average annual escapement, in thousands from 1951 to 1989, of the five major Pacific salmon species for the major sub-catchments in the Fraser River Basin. The numerical importance of the pink and sockeye is evident but equally significant is the variation in areal distribution. The pink are dominant in the Lower Fraser and the sockeye in the upstream catchments. Such geographical variations in abundance are one factor in accounting for the seasonal variations in the patterns of migration. Figure 6.3 illustrates these for the downriver migration of the young and upstream migration of mature adults for pink and sockeye salmon.

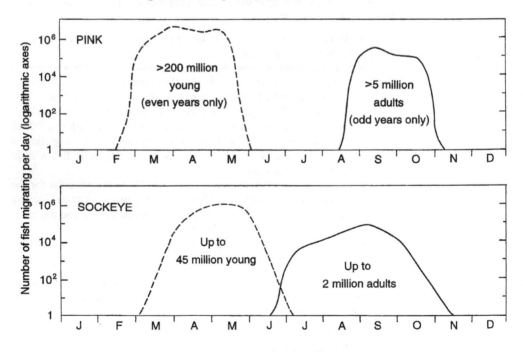

Figure 6.3   Generalised seasonal patterns of pink and sockeye salmon migrations. 'Young' moving downstream, 'adults' moving upstream (Northcote and Burwash 1991)

This basic information on distribution and timing of salmon runs is important in that different species form overlapping populations throughout the Fraser River Basin. Adverse environmental effects, especially those that are site specific, effect different species in different ways. Many forms of salmon fishing are targeted at individual species although deep ocean fishing, by vessels with giant drift nets, does not adequately distinguish between species, salmon or otherwise.

Stakeholders and Conflict

Stakeholders

Conflict over the allocation of the Pacific salmon resource is not new. There is no simple list of stakeholders – the cast is large. Indeed, the cast in the Fraser Basin justifiably merits the term 'hyper-pluralism', used by Amy (1983) to describe an environmental mediation with a plethora of stakeholders who have previously honed their skills to influence decisions. Two key factors in this process were:

- the Pearse Commission in 1982, which suggested the application of environmental economics to the problems; and
- the recognition throughout Canada of changes to the constitutional role of the First Nations.

Stakeholders are by no means limited to local or provincial interests – and this broader interest is not new. Since last century, the federal government has been involved in salmon management on the Fraser. National and international interests are involved, through multinationals such as Alcan (Aluminium Company of Canada), through the increasing globalisation of major environmental issues, and through the rising international concern over the plight and rights of indigenous peoples. The international dimension also emerges through commercial fishing for salmon in international waters and the associated agreements between the United States and Canada, most notably the Pacific Salmon Treaty of 1985. The Canadian federal government has sought to be involved with the salmon fishery since federation in 1867.

It is naive, therefore, to divide fishers of salmon into commercial, aboriginal, recreational and governmental interests without recognising the marked divisions within each group, and the overlap between groups. These divisions are in numbers, salmon species targeted, methods of fishing, industry organisation and in geographical distribution. The battle scars of earlier conflict have left a legacy of alliances and mistrust.

The resolution of conflict in such circumstances requires particular attention to defining the stakeholders. To give order to this process some form of classification is essential.

Aboriginal interests

Prior to colonisation, indigenous Indian bands relied upon fish from the Fraser River, dominantly salmon, for half their diet. Salmon were an integral part of their culture. Throughout the Basin there were similarities in their social organisation, groups were small and settlements rarely exceeded a hundred or so individuals. Each group, or Band, had property rights that defined access to resources of which fish were the key.

Notwithstanding similarities in culture, it is important to note that there are 91 distinct Indian Bands within the Fraser Basin. In 1989 the official Indian population numbered 27,000. This is close to the estimate for 1800, when the adverse effects of introduced diseases had already begun. The population nadir was about 1930, with the total number estimated to be only 8000. The Bands are divided into six language groups, the majority of which are mutually unintelligible.

The Bands are those formally recognised in the Indian Act, each with its own registered membership and leader. Actual numbers are likely to be larger as the official estimates do not include non-Status persons genealogically allied to Band members (mainly women and young children). There are also Indian groups outside the Fraser Basin for whom Pacific salmon play a key cultural role. Examples are the Kwakiutl of North Vancouver Island and the Nuu-chah-nulth of West Vancouver Island.

For millenia the native bands lived in a sustainable relationship with their cousins – the salmon. From colonisation until the last decade their role and voice was virtually ignored. A key factor in the growth of their influence was the international Pacific Salmon Treaty in 1985. This focussed the Canadian government's attention on the need to specify salmon numbers for domestic use, and 400,000 sockeye were allocated to the Indian bands of the Fraser Basin. Dialogue and comanagement became necessary between the Department of Fisheries and Oceans (DFO) and the scattered Indian bands in order to allocate the 400,000. The fledgling Interior Indian Fisheries Commission (IIFC) provided such a body. It immediately received funding and administrative assistance from the DFO. At about this time, the mid-1980s, the Pacific Fishermen's Defense Alliance (PFDA) was formed. Its mission was to oppose the devolution of authority to Native groups. Inevitably an element of racial division was injected into the conflict.

Progressively the role of the IIFC strengthened. The relationship outgrew any lingering paternalistic relationship with the DFO. In 1989 the widely scattered

Indian groups signed a formal treaty codifying objectives, principles and procedures. The treaty is based upon mutual recognition of sovereignty and all
members have agreed that there is a an overriding inter-tribal interest in salmon
conservation. Tribal laws will be subordinate to: 'The principle that Inter-tribal
rights supersede individual tribal rights if the two are in conflict in order to
ensure the survival of the salmon'.

The treaty also specifies the means for achieving consensus and for resolving
any outstanding inter-tribal disputes over fishing policy. Technical studies to
assist with all aspects of salmon management are underway. These are largely
funded by the government but without control of the process. The IIFC was
inspired, in part, by inter-tribal agreements on fishing, many of them for Pacific
salmon, in the north-western states of the USA.

Over a decade, the potential power of the IIFC has transformed the virtually
powerless and voiceless scattered Indian bands into one of the best organised
and recognised stakeholder groups, with national and international links. Currently a framework agreement is in preparation that involves the IIFC's ten tribal
councils, the federal Department of Indian and Northern Affairs, the DFO and the
provincial Ministries of Environment, Agriculture and Fisheries.

Commercial interests
Pink and sockeye salmon stocks from the Fraser River Basin contribute 66 per
cent and 60 per cent respectively of the total British Columbia catch. However,
there is no commercial fishing within the freshwater parts of the Basin. The main
harvesting areas for pink and sockeye are the Johnstone, Georgia and Juan de
Fuca Straits off the west coast of Vancouver Island with some catches in the
Lower Fraser estuary. The DFO recognise ten fishing 'regions' in the area. More
distant waters include the ocean waters to the south of Alaska. The landed
wholesale value of commercially caught salmon from Fraser River stocks in the
late 1980s averaged $270m per year.

The techniques used for commercial fishing are seine nets (which encircle
the fish), usually associated with larger vessels, gill nets (that trap individual
fish) and trolling (on multi-hooked lines). Many of the fleets and workforce are
dispersed into small fishing communities, for instance on the west coast of
Vancouver Island. These include commercial Indian boats which are subject to
the same regulations as non-Indian groups.

Each group has at least one association that promote its interests. The Pacific
Fisheries Alliance is the umbrella organisation for non-Indian commercial
interests. This replaced the earlier Pacific Fishermen's Defense Alliance.

There are also powerful organisations that represent the workers. The largest
of these is the United Fishermens and Allied Workers Union (UFAWU) with
its members drawn from fishing crews and shore workers. The UFAWU is the

primary bargaining agent in negotiations with the processors but is also active in questions of federal fishing policy. For instance, it strongly opposed the Free Trade Agreement with the USA.

The Native Brotherhood of British Columbia (NBCC) is the union for native fishermen. The NBCC has a long history of conflict with the UFAWU. Initially because the NBCC acted as a cover for native property rights and because of divergent actions in support of fishing industry strikes.

The larger processing enterprises are members of the Fisheries Council of British Columbia, traditionally the leading advocacy group for fishing policy. Smaller companies voice their concerns through the Pacific Seafood Council.

Recreational fishing

Three-quarters of the sport fishing catch is from the Straits of Georgia, the bulk of this being salmon. It is estimated that some 300,000 people fish for salmon on a recreational basis. They range from skilled fly fishermen in the upper catchments, to children fishing from wharves in downtown Vancouver, to exclusive lodges and charter boats on the coast. Estimates in the mid-1980s were that fishing within the Basin itself generated $182m of economic output, about half from the Lower Fraser. However, the contribution to the sparsely populated Upper Fraser Basin is a major factor in its economy. It is also clear that angler effort is increasing at a rate that gives concerns as to whether the habitat can meet the demand. The majority of the angler expenditure was from residents of the Province.

Stakeholder groups are numerous, the larger include the Sportfishing Institute, the BC Wildlife Federation and the Sports Fishing Advisory Group, the latter appointed to advise the DFO. In addition, there are countless local fishing clubs.

Environmental groups

These are essentially non-government organisations that vary in size and interest, generally known as NGOs or ENGOs. Salmon conservation is often only one of their concerns and is encompassed by the aims of environmental coalitions which are better able to negotiate with governments. Examples are the Fraser River Coalition, whose focus is on water quality and habitat issues in the Lower Fraser. The Nechako Environmental Coalition was formed to oppose water diversions to meet the energy requirements of Alcan. Here, the interests of native people were very important. Some divergence of objectives always occurs with issues of fish conservation. These are between the more fundamental environmentalists who wish to preserve the ecosystem and fishermen who wish to conserve salmon in order to catch them! Some groups are well connected internationally, for instance, Greenpeace, which was first launched in Vancouver, and is a 'multinational' in its own right.

Government

The governments are involved through their major development corporations and interests as outlined in the section below, and through the maze of agencies, legislation, regulations and policies. These represent conflicts between provincial and federal priorities and between the governments' multiple objectives in terms of balancing environmental protection, indigenous rights, the need for economic development, international obligations, and the overriding imperative of satisfying its various constituencies.

The Canadian federal government has sought to be involved with the salmon fishery since federation in 1867. The British Columbian government is clearly one, if not the, major player. Both levels of government can take very different roles according to the issue, and are often in conflict. In general, the provincial government epitomises a frontier development mentality with a disregard for the rights of indigenous people.

'Those who should be included'

We have seen that the decline in salmon numbers is due to the many groups who directly exploit the resource and to the activities of those who degrade the aquatic environment. The former wish to be included as stakeholders in any forum that seeks to better manage salmon. The latter group can be regarded as reluctant stakeholders, but for effective resolution of the conflict it is essential that they participate.

The 'those who should be included' stakeholders are dominated by large organisations, a mix of government agencies and large corporations. They reflect the traditionally strong resource development ethic of western Canada. Despite rewritten corporate statements, for many of them the effects of their development on salmon is still a fringe issue. This extends to relationships with the First Nations. The property rights of the Indians have progressively been upheld by the courts but it is disturbing that the judicial decisions have been more slowly recognised in the rough and tumble of political reality.

Foremost among these are the landholders whose activities (the distribution of waste water discharges is shown in Figure 6.4) impinge directly on the rivers. Four key groups can be recognised:

- Forestry
- Energy
- Transport
- Local governments

The conflicts caused by the different stakeholders are presented in more detail in the following section – Conflicts.

- **Forestry:** Over 90 per cent of land in British Columbia is owned by the provincial government: this includes virtually all forest lands in the Fraser Basin. The argument is, therefore, that the location and controls on present and future logging are the responsibility of the Ministry of Forests, thus measures to reduce sediment discharge are in the hands of the provincial government. The spoilation of gravel spawning streams by sedimentation is likely to remain a problem in the upper basins, its effects on sediment yields in the lower river will be small. The effluent from major pulp mills in the upper catchment poses a similar challenge (Figure 6.4).
- **Energy:** In BC the generation of energy equates with hydropower. In 1962, the provincial government created a public utility, the BC Hydro and Power Authority, to coordinate the development and distribution of electricity. For many years this was seen as an engineering dominated development agency. Modification to the role of the 'Hydro' now encompasses 'minimising adverse effects on the natural and social environment'. By Canadian standards the Fraser basin, with the exception of the Keman Project on the Nechako River, is not a major producer of hydroelectricity.
- **Transport:** The major effects of transport relate to the engineering of rail tracks and roads. The terrain limits these to the deep glaciated river valleys that provide the only routes through the mountain ranges that separate BC from eastern Canada. A major example of the problems that have arisen was the landslides at Hell's Gate, some 200 km upstream from Vancouver on the Fraser River. These occurred in 1913 and caused major blockages to the salmon migration routes. The prospects of similar problems from the twin tracking of the Canadian National rail lines in the mid-1980s created major legal confrontation.
- **Local governments:** The major concern is the disposal of poorly treated sewage, storm water and landfill effluent into the river system (see Figure 6.4). It is noteworthy that the smaller upstream councils have markedly better effluent treatment than the larger councils of Greater Vancouver.

Conflicts

The conflicts surrounding salmon fall into two general groups: those directly concerning the fish catch; and those involving activities which affect some part of the salmon's habitat and therefore indirectly affect the fish catch. The major areas of conflict are:

- fish catch: the hunted vs the hunters;
- fishers: aboriginal vs non-aboriginal fishing;
- loss of salmon habitat: salmon vs agricultural land drainage;
- interferences with river flow: salmon vs railway and hydro;
- water quality changes due to sediment load: salmon vs logging;
- water quality changes due to solutes: salmon vs pulp mill and municipal runoff.

It is important to appreciate that a 'frontier' mentality still exists in many parts of BC; and that this is associated with a strong pro-development attitude.

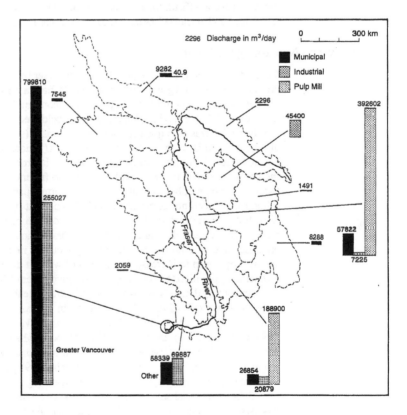

Figure 6.4 Daily municipal, industrial and pulp mill discharges in the Fraser River sub-basins (in m^3/day) (Hall, Schreier and Brown 1991)

Overfishing: Salmon versus the Hunters

Salmon stocks are today about one-fifth of their numbers a century ago. However, this dramatic reduction still leaves a valuable fishery. A major reduction in fish stocks, and increasing catch effort for a heavily capitalised and electronically sophisticated commercial fleet, are unfortunately characteristics of most major fisheries the world over.

Operational concepts such as 'maximum sustainable yield' may leave the stocks very vulnerable. Fish populations are very dynamic and may vary greatly from year to year. In a year when the population may be low due to natural or other factors, the fish may have a difficult time evading capture. The worst case scenario is that the remaining numbers fall below the critical value to ensure continuation into the future.

Aboriginal versus Non-aboriginal Fishing

Conflicts between the native peoples of the Fraser basin, the later arrivals and provincial and federal governments of Canada, have a very long history. Salmon has been a major, but not the only, source of conflict. Land lost to the Kenney reservoir which dammed the Nechako River belonged to the Cheslatta Indians. The Nechako was a major tributary of the Fraser (see Figure 6.2). The dam diverts two-thirds of the flow from some 5000 square miles of catchment to Alcan's (the Aluminium Company of Canada) hydroelectric power plant. The Indians were forced to move at short notice to fragmented land holdings. The focus was lost land, but they also lost an important part of their livelihood: the salmon which can no longer use the Nechako. That was in 1952 and protests were few and ineffective. But over the last decade or so, native issues have received an increasingly higher profile. The First Nations have become very well organised and determined and have many victories in court. They are battling attempts by Alcan to increase the amount of water it diverts from the Nechako – water long promised to Alcan by the BC government (Boyer, 1986). They are also confronting the railways over proposals to double the track along the Fraser.

Some 96 per cent of the Indian catch within the basin consists of sockeye (85 per cent) and pink (11 per cent) salmon. This has shown a steady growth since the 1950s (see Figure 6.5). Since 1985, 400,000 sockeye have been allocated for the use of status Indians. Nevertheless, the Indians depend on salmon that run the gauntlet of the maritime commercial fleets which take 90 per cent of the total catch.

A growing conflict with non-Indian groups is the degree to which the Indian catch is used for commercial sale as opposed to food or ceremony. Progressively Indian spokespersons are arguing a case for commercial sale. (The native Brotherhood of British Columbia represents native commercial fishing interests.) This is seen as a continuation of the trading of fish, part of the traditional culture of the First Nations.

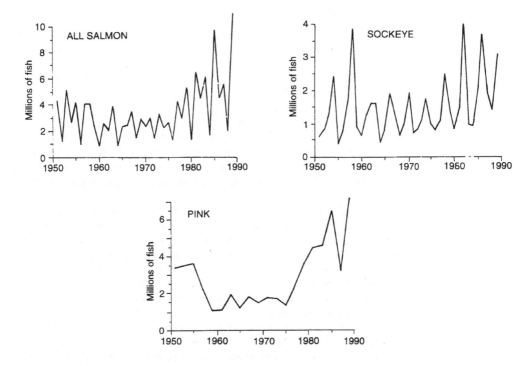

Figure 6.5   Annual native Indian fish catch for all salmon and for sockeye and pink salmon
(Northcote and Burwash 1991)

Loss of Habitat: Salmon versus Agricultural Land Drainage

Modifications to the biophysical environment are basin-wide but the structural works on the floodplains of the Lower Fraser represent a major threat. *Environment Canada*, a federal government agency, has estimated that 80 per cent of the Fraser wetlands, including the inter-tidal and sub-tidal zones, have been drained and converted to other uses, primarily agriculture. This has greatly decreased the habitat available for rearing and as a refuge from predators. Figure 6.2 shows that such changes to the main stem of the Fraser and to the estuary effect the migration of all species. Urbanisation of the lower river, especially for the Vancouver conurbation, has led to armouring of the river banks, channelisation and associated dredging all of which impinge on spawning and migration.

91

Changes in the River Regime: Salmon versus Railways and Hydroelectricity

Changes to the natural river regime occur at all scales. Historically, the rockslides and rock dumping into the river at Hell's Gate had a major impact. Excavations and blasting associated with railway construction in the period 1912–1915 caused massive slides which blocked the river at Hell's Gate, on the lower reaches of the river. The rockslide in 1913 coincided with the four-year peak of the sockeye salmon cycle. This prevented a peak year of sockeye from reaching upstream spawning grounds and eliminated for decades the large upstream stocks of pink salmon – all those who fish for salmon were affected, but none worse than the upstream Indians. It was not until the 1940s that fishways were constructed to allow salmon to move reasonably easily past Hell's Gate, although the blockage still poses problems in low flow years.

The Hell's Gate catastrophe of 1913 and construction of fishways around the blockage did not signal the end of conflict between transport construction and salmon. The issue re-emerged in the mid-1980s with the proposal by the Canadian National Railways (CNR) to twin-track the route in the Fraser–Thompson corridor. This caused encroachment on land of the Oregon Jack Creek Indian band, whose chief was Robert Pasco, and interference with fishing. The project was unusual in that it was designated by the federal government as of major public importance and was not subject to obstruction by government departments. An environmental impact assessment panel was established and Chief Pasco was a member. However, before the report was presented, CNR announced that it would go ahead with the scheme. Pasco resigned from the panel and, despite threats of blockades of the track, no federal or government agency took up the cause of the Indian band. The only avenue left was to take out a court injunction to stop the work proceeding. This was successful and the final judicial outcome is awaited.

It is worth observing that some Indians had suggested a way around the confrontation. They suggested combining traffic from the two railway companies and using their separate tracks to achieve the same effect as twinning – at a considerable cost saving: thus all trains on one track would run downstream and on the other upstream (Boyer, 1986).

The major Fraser River is not dammed in the conventional sense, although for a time the Hell's Gate blockage had similar effect. The possibility was, however, raised seriously in the 1950s. The major reason for this apparent forbearance is the need to maintain the river for the annual migration of approximately 3.5 million salmon.

There are hundreds of dams scattered throughout the Basin, many designed for the generation of hydro-power at the local scale. All cause major modifications to flow and barriers to upstream migration as well as direct mortality of juvenile fish as they move through the turbines. The sockeye, huge numbers of which spawn in the upper tributaries of the Fraser system (see Figure 6.2), are particularly prone to such effects. Major water diversions in the Nechaka sub-basin have diverted some 67 per cent of that river's discharge for hydro-power developed by the Alcan. This diversion has not only reduced salmon numbers in the Nechaka but in the downstream Middle Fraser.

Stakeholders interested in fish conservation have expressed alarm at the Hydros' current encouragement of small scale development based on private operators. Many of these are less likely to be built and maintained to the same standards as larger plants – for example in terms of the provision of fish ladders and adequate water flow for their effective use. Governments are multi-objective; they need to generate electricity, but also to maintain fisheries, satisfy the environmental lobby and when appropriate honour their obligations under international law. Small hydroelectric companies will generally have a single objective: to generate and sell electricity at the lowest production cost. The resources of ENGOs would not be capable of monitoring numerous small developments, let alone checking on their likely environmental impacts and if necessary opposing the development.

Water Quality – Sediment Load: Salmon versus Logging

The recent glacial history of the Fraser Basin has bequeathed thousands of kilometres of clear, gravel-based watercourses. Any interference, especially with the steep slopes of the majority of the catchments, poses a risk of soil erosion and enhanced sediment load for the Fraser and its tributaries. Logging and associated roading are the major source of sediment runoff, much of which is composed of the finer fraction that causes siltation of the gravel beds. This reduces benthic food habitats and infills the submerged gravel spacing which is critical for salmon overwintering. In some areas it has been shown that siltation decreases the fry emergence by some 30 per cent.

The problems posed by future logging are likely to exacerbate siltation. This is because only 16 per cent of forests in the coastal sub-catchments are mature compared to some 50 per cent in the upper catchments. This will lead to pressure for logging in the upper catchments which have steeper slopes and slower regeneration.

Siltation is the most critical of the problems due to forest activity but clear felling also causes increases in water temperature. Other problems relate to log booms, especially in the Fraser estuary where some 1500 hectares are leased for boom storage.

Water Quality – Solutes: Salmon versus Pulp Mills and Municipal Sewage and Runoff

There are marked regional variations in the type and volumes of effluents discharged into the Fraser river system. There are five major pulp mills in the middle and upper Fraser, with an average combined daily discharge of 600,000 m^3 (cubic metres) per day. Municipal average daily discharge is over 800,000 m^3 per day, this is overwhelmingly from the Lower Fraser River and its estuary. The broad pattern of these discharges is illustrated in Figure 6.4. These data do not include the stormwater runoff from the Greater Vancouver region.

Recent years have seen widespread community concern at the potential toxicity of pulp mill discharge, particularly by chlorinated organic compounds. In response to this pressure, the federal government has undertaken detailed studies of pulp mill effluent. Studies of the pulp mills in the Fraser Basin are interpreted as indicating relatively limited effects on fish stocks mainly due to the rapid dilution of flows downstream. Standards have been tightened in response to community concern.

Well over 90 per cent of the municipal effluent discharged into the lower estuary is either untreated or treated only to primary standard. The much smaller volumes discharged by the municipalities in the middle and upper basins are generally treated to a much higher standard. This is most likely to be a reflection of the long-held view of the world's coastal cities that discharging poorly treated sewage into estuaries is acceptable. To this estuarine municipal effluent must be added huge volumes of storm water runoff and an estimated 7500 m^3 of landfill leachate per day, high in organics and ammonia.

There is no doubt that these effluents have an adverse effect on aquatic organisms including salmon. The nature of the links and the reduction of salmon numbers is, however, still unclear. In the upper basin there are local problems such as placer mining, including residual mercury from early forms of processing.

Over-exploitation and deleterious effects to the aquatic environment have seriously depleted fish stocks. However, there is some good news in that there is a small but statistically significant increase in numbers of Pacific salmon since the early 1950s. This is particularly the case for pink salmon, although not statistically significant for sockeye. The cyclic nature of the salmon numbers is also clear.

Conflict Management/Resolution

Many approaches have been used to deal with the conflicts surrounding salmon in the Fraser River basin. These range from highly adversarial and confrontational techniques to the gentle persuasion of incremental change through evolving social values expressed through politicians and the judiciary. Here, some of the mechanisms used are summarised briefly under four headings:

- Bureaucratic
- Economic
- Political judicial
- Community.

Negotiation and mediation based techniques have roles and are used within all five categories. Although we have concentrated on the formal documented mechanisms, informal negotiation is usually widespread and often effective.

The Bureaucratic Approach

Bureaucratic mechanisms include all the ways in which conflicts are managed and resolved within government administrations. These include the various leases, licences and permit procedures and associated resource management guidelines; as well as task forces, impact assessment procedures and planning processes. There is a bureaucratic jungle with a maze of overlapping agency boundaries.

Historically, the DFO was the arbitrator of equity on behalf of the Canadian government. Its strength was its scientific expertise, its conservation ethic and the ultimate decision-making power of the minister. It has progressively espoused policies of 'co-management' and 'consensus', although it finds problems in relinquishing its supremacy and its conservation goals, and has been under pressure to use economic policy instruments. The Pearse Commission report in 1982, *Turning the Tide: a New Policy for Canada's Pacific Fisheries*, questioned the conservation role of the DFO and replaced it with the concepts of 'maximum sustainable yield' and 'limited entry' to the salmon fishery. The Commission hearings also acted as the focus to orchestrate new voices.

Compliance with regulations has relied almost entirely on the threat of prosecutions. Until recently these have been minimal and there has been a general lack of enforcement. However, this is changing and enforcement has become more vigorous.

The Pacific Salmon Treaty incorporates pre-season and in-season negotiations for fish allocation and management. The Fraser Panel is a component of the organisational arrangements.

Use of Market Forces

Essentially, this mechanism uses market forces by allowing prices to balance supply and demand. This may involve the creation of markets; or involvement by government to ensure that markets function properly – for example, to ensure that the beneficiary of a development bears all the costs rather than displacing them to another resource user or to the government.

A prime example is the 'no net loss' principle, a key policy outlined by the DFO in late 1986. Ideally the assessment of a new development should avoid habitat loss. Where this is not possible the replacement of lost habitat at, or near, the site is the next preferred option. If such measures are impossible, compensation is required to make good habitat loss by other means. Such measures could be the building of a hatchery to replace lost natural stocks, the costs borne by the proponent of the development. The initial application of the 'no net loss' principle was in the Fraser River Estuary as a component of The Fraser River Estuary Management Program (FREMP). FREMP has been hailed as a world class model of environmental economy partnership. Essentially this is a performance driven approach – the performance standard is 'no net loss', the methods of achieving this are negotiable. An obvious and critical question concerns the long-term viability of some of the methods.

The Pearse Commission argued for the use of environmental economics in the management of the Fraser River.

The Judicial Approach

This includes all Canadian courts from the Supreme Court of Canada to the local community courts. Conflicts are resolved through administration and interpretation of legislation and common law. Occasionally, courts may appear to make law when interpretation in the light of different criteria, changing community values, or new international obligations, leads to apparently radical results. Until recently there has been relatively little use of the courts to resolve water resource conflicts with the outstanding exception of the Privy Council cases mentioned below. Changes to the laws of standing and changes in the attitudes of the courts to native people's rights, among other things, have led to a greater willingness to pursue solutions though the legal system.

Conflict over the federal/provincial control of fisheries was considered by the Privy Council in London three times between 1898 and 1920. The outcome was that the federal government retained the right to regulate anadromous species (including all Pacific salmon species) and the province manages other forms of inland fisheries.

In 1973 the Supreme Court of Canada held that aboriginal rights were part of the law of the land. In the context of the use of the salmon resource in the Fraser Basin definitions are still awaited. Nevertheless, a 1990 Supreme Court of Canada decision concerning Sparrow, a Musqueam Indian, was a landmark. The court upheld Sparrow's aboriginal right to fish for food, social and ceremonial purposes notwithstanding the licence specifications of the Fisheries Act.

Quasi judicial
The Pearse Commission is the outstanding example of a quasi judicial body with a conflict management mandate.

Political Mechanisms

Political mechanisms include all the legislative bodies established to manage and resolve conflicts through the decisions of elected representatives. It also includes actions of the executive acting independently of the legislature.

The BC government has a strong development ethos, and refuses to recognise distinct indigenous people's rights. This has occasionally put it on a collision course with native people, environmental interests and the federal government – for instance, over indigenous rights, where it has a poor record in the courts. It took the stance that the activities of the (federal) Environmental Protection Authority were an intrusion into provincial jurisdiction.

The recession of the 1980s and associated fiscal austerity led to more than the usual government emphasis on major resource and infrastructure development, and naturally to some major conflicts. But this has not been confined to the provincial government. The federal government attempted to 'fast track' the major railway project which would have led to increased sedimentation and other problems for salmon in the Fraser. Following court action by objectors, the issue is still unresolved.

Community Interests

Community interests include recreational fishers, indigenous peoples, industrial interests and conservation groups. Over the past decade, each has found a stronger voice. The protagonists have each formed a formal lobbying organisation, in order to attract the attention of the political, bureaucratic and judicial aims of decision making and to make maximum use of the media. One outcome is that everyone at the local, provincial, national and international levels of decision making are more aware of the issues. Another outcome is that the competing arguments become more oppositional and stylised; and the voting public more confused about the issues. Meanwhile the Indian fish catch rises (Figure 6.5).

Table 6.3   Evaluation based on the old and new rules

• **Exhaustable v finite**	Canada is to be congratulated for its early recognition, in the final decades of the 1800s, that Pacific salmon were not an inexhaustible resource. The *Water Acts* recognised this and led to the formation of the Dominion Department of Fisheries, the forerunner of the DFO, to manage the resource. International agreements with the USA addressed the need to control ocean fisheries – essential for a migratory resource that did not stand still to be counted! We can only conjecture on the fate of this rich natural legacy if these early responses had not occurred. However, it is a fact that fish numbers declined by 80 per cent from the turn of the century to the early 1950s.
• **Sub-system or Ecosystem**	DFO is a world leader in fish biology but it is clear that the depletion of Pacific salmon stocks require a wider consideration of the ecosystem. It is impossible to separate land and water systems or biophysical and socio-economic influences. Sediment generated from logging poses a major threat to the gravel spawning beds of the rivers, effluent from pulp mills, mining and urban centres represent issues that extend beyond the aquatic ecosystem.
• **Social values do change**	Until about 1800, salmon and native peoples enjoyed a sustainable relationship. Social and cultural values were in balance with biophysical systems. This underwent cataclysmic change. The social values were those of colonial exploitation of the resource. Legislation and the formation of the DFO made federal government the arbiter of environmental and social values. The new social order has begun to recog-

nise the rights of aboriginal peoples, the recreational needs of all residents and the quality of life. Multi-purpose but different.

- **Environmental management is a people matter**

Observers of the salmon conflict opine that the biological research and fish management have a worldwide reputation. In stark contact, there is little evidence that the DFO or other stakeholders have made any effort to understand their own or other organizations' dynamics. The conflicts, and the solutions required to solve them, are overwhelmingly people problems.

- **Authoritarian or consensual decision making?**

Authoritarian decision-making processes still predominate, despite much greater attention to public consultation and involvement. The approach to the railway dispute epitomises an authoritarian approach. However, there appears to be much greater use of informal negotiation in the implementation of water resource management than is generally recognised.

- **Technology – limits and costs**

The goal of the DFO has been to increase fish numbers, the strategy has been to improve our understanding of the fish and their aquatic habitat. This has involved the expenditure of large sums over many decades. Does this concept of sustainability match that espoused by resource economists of maximum sustainable yield? The significant influences on the habitat are from exploitation and from a range of deleterious factors associated with other forms of development. Logging, transport construction, pulp mills and sewage effluent affect fish populations, however advanced the technologies to reduce the effects they will not provide the final answers: they nibble at the symptoms.

Conclusion

The history of the Fraser River salmon resource is a microcosm of renewable natural resources at a world scale. The story abounds with one and a half centuries of conflicts. These have increased in complexity and contain a growing number of stakeholders. Tony Dorcey (1991 pp 282) concludes his comprehensive two-volume study of the Fraser River Basin with the words:

> Reaching agreement despite conflicting values and uncertain science is the greatest challenge we face in the sustainable development of the Fraser River Basin.

This gives emphasis to the theme of our volume that conflict and opportunity are linked.

Dorcey continues:

> There are good reasons for optimism about the future of the Fraser River Basin, particularly when it is compared with the situation in many other river basins in the world. Overall the Basin still has a relatively low level of development and the environmental problems created by past activities are to a large extent reversible.

This matches our analogy in Chapter 1 that the natural environment vis-à-vis resource development can be regarded as 'half full' or 'half empty', depending on whether the situation is being viewed from an optimistic or pessimistic viewpoint.

Pessimism for the Fraser salmon lies in:

- the 80 per cent decline in salmon numbers since the turn of the century;
- adverse changes to the river environment;
- the overuse of the resource by a range of fishing interests – commercial, recreational and First Nation;
- the increasing complexity of the bureaucratic framework;
- the ingrained pioneering ethic of British Columbia with its emphasis on natural resource development, and the apparent willingness of government to support major infrastructure and industrial projects at the expense of the natural environment and salmon fishery.

Optimism would stress:

- the maintenance of salmon stocks and catches since about 1950;
- the recognition of the rights of the First Nations with their overwhelming interest in maintaining the salmon stocks and their habitat;
- the growing environmental ethic with government recognition of the fragility of the resource and controls on all forms of pollution;

- the increased awareness that the sustainability of the salmon is a multi-faceted problem that can only be achieved within a truly transdisciplinary approach;
- the recognition that conflict analysis and management has a vital role if the problems are to be resolved.

Sustainable development is currently seen as the key to future resource development. For success, this involves awareness of the problem, the recognition that there are major conflicts and the need for all stakeholders to reach accord on their determination to achieve the best win-win situation for the environment and the community – present and future.

The growing internationalisation and associated complexity of major environmental problems is a double-edged sword. It has introduced and gained widespread acceptance in principle of the concept of sustainable development. Indigenous peoples, environmental interests and even local communities are increasingly well networked with similar interests in other parts of the world – this is not to suggest that the interests of these different groups necessarily coincide, although in general they would all be interested in maintaining the salmon. Multinational industries are also improving their ability to mobilise public opinion in their support. The existence of 'free trade' agreements is generally seen as very unfortunate for the natural environment and for attempts by governments to regulate solely on environmental grounds. To some extent this may be countered by international agreements covering specific species such as salmon.

Whether the future for the Fraser salmon is one of hope and optimism or despair and pessimism remains to be seen. The chapter title 'A heroic effort' comes from a fish biologist. It reflects the hopes for the future from the highly advanced biological management of the salmon. Its optimism ignores the unresolved pressures on the salmon population from industrial development, unsympathetic administration and competing harvesters.

References and Further Reading

Amy, D J (1983) 'Environmental mediation: an alternative approach to policy stalemates' *Journal of Policy Sciences* vol 14 pp 345–65

Boyer, D S (1986) 'The untamed Fraser River' *National Geographic*, July pp 45–75

Dale, M G (1991) 'The quest for consensus on sustainable development in the use and management of Fraser River salmon' in Dorcey (Ed) (1991a) *Perspectives on Sustainable Development in Water Management: Towards Agreement in the Fraser River Basin* Westwater, University of British Columbia, pp 155–88

DFO (1989) *Pacific region strategy outlook. Vision 2000. A vision of Pacific fisheries at the beginning of the century. Discussion Draft* Canadian Department of Fisheries and Oceans

Dorcey, A H J (1991) 'Water in sustainable development of the Fraser River Basin' in Dorcey (Ed) (1991b) *Water in Sustainable Development: Exploring our Common Future in the Fraser River Basin* Westwater, University of British Columbia, pp 3–17

Dorcey, A H J (1991) 'Sustaining the Greater Fraser River Basin' in Dorcey (Ed) (1991b) *Water in Sustainable Development: Exploring our Common Future in the Fraser River Basin* Westwater, University of British Columbia, pp 269–83

Hall, K J, Schreier, H and Brown, S J (1991) 'Water quality in the Fraser River Basin' in Dorcey (Ed) (1991b) *Water in Sustainable Development: Exploring our Common Future in the Fraser River Basin* Westwater, University of British Columbia, pp 41–71

Kew, J E M and Griggs, J R (1991) 'Native Indians of the Fraser Basin: towards a model of sustainability' in Dorcey (Ed) (1991a) *Perspectives on Sustainable Development in Water Management: Towards Agreement in the Fraser River Basin* Westwater, University of British Columbia, pp 17–48

Northcote, T G and Burwash, M D (1991) 'Fish and fish habitats of the Fraser River Basin' in Dorcey (Ed) (1991b) *Water in Sustainable Development: Exploring our Common Future in the Fraser River Basin* Westwater, University of British Columbia, pp 117–38

Thompson, A R (1991) 'Aboriginal rights and sustainable development in the Fraser-Thompson Corridor' in Dorcey (Ed) (1991a) *Perspectives on Sustainable Development in Water Management: Towards Agreement in the Fraser River Basin* Westwater, University of British Columbia, pp 473–518

WECD (1987) *Our common future* World Commission on Environment and Development, Oxford University Press, Oxford

7 Saving a Shire:

Reclaiming a salt-affected irrigation area in Victoria*

Bruce McKenzie, Linden Orr and Peter Crabb

INTRODUCTION

The Nyah Irrigation District is located on the Victorian banks of the River Murray, approximately 30 km north-west of Swan Hill. The District covers an area of some 1560 ha and includes the townships of Nyah and Vinifera. The district has a population of some 1800, of which 530 reside in Nyah West and 360 in Nyah. In terms of overall numbers, the population is more or less static. However, there are indications that it is an ageing population, with 28 per cent of both the total population and the number of landholders being over 55 years of age.

This case study of the Nyah Irrigation District follows the structure outlined in Part 1 of this volume. Section 1 of this chapter provides background information relating to **why** environmental management in the District is a conflict. Section 2 provides the reader with a wide range of material necessary

* This case study is based very largely on the *Nyah Irrigation District Status Report* produced by the Nyah District Action Group (1991). The authors would like to thank all of those who helped them prepare this case study. In particular, the authors benefited considerably from discussions with and insights provided by Ken McDougall, the Community Development Officer working with the Nyah District Action Group.

for the exploration and analysis stages of environmental decision-making, ie **what** is being managed in the area. Section 3 outlines how the Shire is currently managed. Section 4 is concerned with **how** the District is to be managed in the future and in particular the critical community involvement in this. This represents the goalsetting stage in environmental decision making and identifies strategies being utilised in terms of 'Conflict Management – Scaling the Mountain' (Figure 4.2). Section 5 lists the key players, the **who**, in the conflict management process.

SECTION 1: WHY ENVIRONMENTAL MANAGEMENT FOR NYAH SHIRE?

The Conflicts

The history of Nyah indicates that continued concern for its economic, environmental and social well-being have been major issues since its early development.

The most recent period of active community concern for the well-being of the District had its origins in a horticultural census undertaken in 1983 (SCHDA 1987). This census confirmed some of the fears of both the Victorian Rural Water Commission and the Department of Agriculture, namely that the area of horticultural land was shrinking, that modern techniques were not being readily adopted to increase production, and that the available irrigation infrastructure was not being utilised.

With the census information in hand, the concerned government agencies invited representatives from the community, both horticulturalists and business people, to a meeting. Approximately 20 interested people attended and at the end of the gathering, it was decided to call a public meeting to form a group to act on the problems. Unfortunately, only about 20 people attended the public meeting. Because of the poor response from the community, it was felt that there was not sufficient support to form an action group.

Nonetheless, the Department of Agriculture and the Rural Water Commission's Irrigation Services Branch embarked on an ambitious extension exercise. Agricultural staff worked in the horticultural industry in the Sunraysia district for a short period of time to gain experience and knowledge which could be used to help Nyah farmers increase their production. The Irrigation Services Branch worked on irrigation scheduling and plant requirements. However, overall horticultural production continued to decline. Several farmers adopted the new methods, replanted with new rootstocks and improved their irrigation management, but overall, the area lost to horticulture increased. The downturn of the dried fruits industry and the Vine Pull Scheme of the mid-eighties saw a further

20 per cent reduction in the area under vines.

In 1988, a sociological survey indicated the main problems perceived by farmers in the district to be small farm size, the reduction of the land in horticulture, high turnover of farms and the lack of experienced farmers. The survey also recognised that there was a lack of direction from Nyah District community to address the continuing downward spiral.

Today, the areas of conflict confronting the community are:

- the non-viability of current block sizes;
- local vs national perspectives on direction for change;
- alienation of horticultural land by hobby farmers;
- the cost of maintaining a water entitlement when it is not used;
- adopting technology to improve water usage efficiency does not result in a cost reduction to the irrigator;
- saline and nutrient discharges from horticultural land to river flats and forest areas;
- planning controls which prevent houses on small blocks being sold as separate titles to allow amalgamation of horticultural land;
- selection of economically and environmentally sustainable crops;
- prioritising of resources for structural adjustment; and
- differences in valuing the use of land, eg economic vs ecological.

If one issue was to be picked out as the most prominent, it would be the small size and non-viability of the land holdings, certainly when compared with other nearby irrigation districts (Table 7.1). This finds further support in the fact that only 14 per cent of the district's farmers are full-time, with only a small minority of these gaining all or most of their income from their holdings.

Table 7.1 Critical factors relating to farming in the Nyah Irrigation District and the adjoining districts of Tresco and Woorinen, 1990

	Nyah	Tresco	Woorinen
Number of farms	200	130	170
Average farm size (ha)	8	9	10
Total value of production (in $ millions)	2	4	9
Average farm income (in $)	5 000	13 000	17 500
Gross margin per hectare (in $)	1000	2 000	2 500

Source: Paper prepared by the Department of Agriculture Victoria for the Murray Darling Basin Structural Adjustment Working Group.

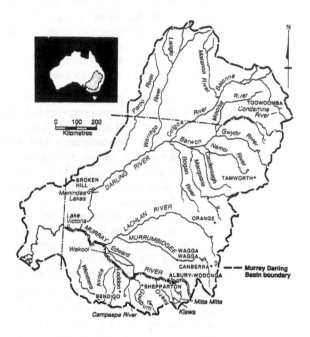

Figure 7.1   The Murray-Darling Basin

A History of the Nyah Irrigation Settlement

Throughout the history of Nyah, the **old** and **new** rules of Environmental Management, Tables 2.1 and 2.2 (pages 36 and 38) may now be observed with the advantage of hindsight. As this short review makes clear, the old rules directed action and the new rules, although present, were ignored. The rules are noted at various points in the text. The history also recounts many periods of conflict, most of which were side-stepped rather than being confronted and managed. The conflicts as they relate to Table 2.3 are also noted in the text.

The Story (Barr and Cary 1990)

Australia has experienced many periods of 'boom and bust'. Such periods have always affected both the entrepreneurs and the working people, and unemployment has always been an inevitable consequence. Over the years, many solutions, some very idealistic, have been proposed to cope with the unemployment. The Tucker Village Settlement Association, formed by a Melbourne church group, proposed the creation of village settlements across the countryside as an alternative to the mass unemployment and poverty of the cities. The dream was of cooperative communities of God-fearing part-time farmers and labourers. To create these communities, land would be subdivided into blocks, too small to allow settlers to develop into full-time farmers, but large enough to provide protection against depression unemployment. The settlements would govern and develop as cooperatives.

The Tucker Association lobbied the Victorian Government for legislation to allow village settlements to be created and in 1893, a Land Act was passed permitting such settlements and providing a government subsidy to assist settlers to establish themselves on their blocks. That same year a village was created on the banks of the Murray about 30 km north-west of Swan Hill, a location that had already been established as an irrigation area in 1910 (SRWSC 1983, pp 2-3). The founders expected the village to develop using the nearby river for irrigation water. The largest block size was just over 20 hectares, quite appropriate with irrigation water. Without water, it would have been woefully small compared with the 260 hectares selection blocks in the adjoining dryland Mallee country (*Old Rule 1: water, air and soil are inexhaustible resources*).

Soon after the first settlers arrived, a nursery began growing apples and vine stock for settlers to plant their orchards. A minor difficulty was that the settlement lay on sand dunes 20 metres above the river. A pump was needed (*Old Rule 5: technology can make anything possible*). The cooperatives' funds stretched to a pump, but nothing was left to pipe the water to the settlement. The settlers hoped the government would pay for this. In the best traditions of rural Victoria, the settlement was named after the politician Taverner, who it was hoped would assist in getting this government assistance. But the rhetoric of cooperative management was discredited in Melbourne (*Conflict Issue 1: inevitable part of change*). The government had lost a lot of taxpayers' money in failed cooperative water projects elsewhere in the colony. Individual farmers were more likely to get help than cooperating groups. The pump quietly rusted. The villagers turned to local government for another pump to get stock and domestic water up from the river. Swan Hill Shire put one in, but it was inadequate to the task and soon failed. By the Federation drought of 1901, villagers were carting water up from the river in drays (*New Rule 5: technology has limits and costs*).

With such a pathetic water supply, dreams of orchards and vineyards evaporated. The villagers grew hay lucerne on their small plots and changed the name of their settlement from Taverner to Nyah, in recognition of the failure of flattery. As the founders had envisaged, the settlers could not support themselves on the proceeds of the small blocks alone. The villagers needed off-farm work to provide for their families. Unfortunately, the Tucker Association idealists had overlooked another problem. There were no local employers. Villagers were forced to travel great distances for work and stay away from their farms for long periods. Many properties were abandoned to the rabbits. The village dream foundered.

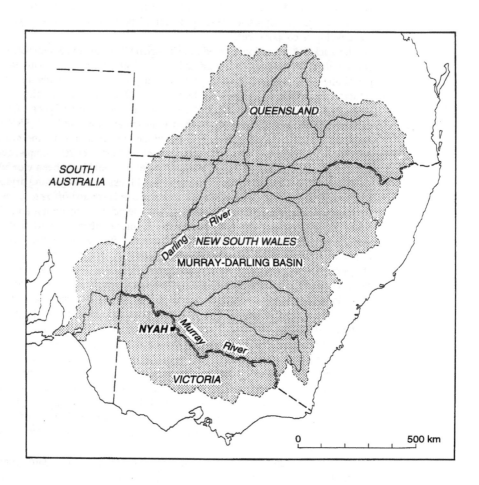

Figure 7.2   The location of Nyah in the Murray-Darling Basin

The Seeds of the Problem: Closer Settlement

The decline of Nyah coincided with the regrowth of government interest in closer settlement. By 1910 the newly created State Rivers and Water Supply Commission was looking for sites to develop as closer settlement schemes. The remaining Nyah settlers interested the Commission in Nyah as a suitable sight for development. In 1911, much that the early villagers had dreamed of appeared as part of the transformation of Nyah into a closer settlement scheme: pump, pipes and channels. Optimism and enthusiasm amongst the remaining villagers was high. A reporter for the local Swan Hill newspaper, J E Robertson (1912), expressed their hopes in untempered enthusiasm:

> Nyah is considered by many people best able to judge to be a second Mildura, and rightly so on account of the potentialities of its rich soil.... As an irrigation settlement Nyah has natural advantages which would be hard to surpass in any part of Victoria, owing to its wealth of soil resource and its excellent geographical position for despatching its produce by river, soon by rail.... One is favourably impressed on a visit to Nyah with the rapid strides it has made in the past two years. The vines and fruit trees in well laid out and beautiful orchards present a pleasing spectacle, predicting the ultimate prosperity which the landowners will enjoy.

Time was to eventually unmask most of the naive reporter's predictions as false. His propaganda was obviously written with an eye to what the locals wanted to read. It did not convince many outside Nyah. By 1914, new settlers had purchased only three of the 40 Nyah closer settlement blocks. The few Nyah farmers watched the slow progress of their settlement with apprehension. The unoccupied blocks proved a financial burden on both the government and the settlers. Nyah farmers vigorously opposed the development of other irrigation settlements in their area while blocks remained unoccupied at Nyah. However, the State Rivers and Water Supply Commission saw no reason for restraint. It developed the nearby settlement of Woorinen in 1914. Nyah found itself in competition with Woorinen for the very limited pool of settlers. Woorinen was closer to the rural city of Swan Hill, and the terms of purchase were more attractive than those available at Nyah. To make matters worse, the Commission enticed experienced settlers from downstream Mildura to Woorinen and used their presence to advertise the success of the area (Scholes 1989 p 119). There was no such public relations stunt to help Nyah.

To compound matters, the State Government's promise of a railway station was only partly fulfilled. A coalition of wheat farmers with legitimate interests in lower freight costs and land speculators with less legitimate interests, lobbied for the station to be sited inland from Nyah. The new station was built a few miles from Nyah, suspiciously next to an already planned private town subdivision. The new town was named Nyah Rail (later Nyah West). Nyah Rail grew quickly, and soon began to dominate its languishing neighbour Nyah. Ill-feeling and tension between the two towns continued for generations, ironic for a settlement originally founded on dreams of cooperative village living.

Expediency and Allocating the Blame: Soldier Settlement

It took the First World War to fill the Nyah blocks. When the enthusiasm for soldier settlements gathered pace in 1916, there was Nyah, an already developed settlement with too few settlers. Nyah was expanded with two new areas, Irrigatia and Nyah Extension, and the government filled the blocks with returned soldiers.

The inexperienced soldier settlers had little cause to rejoice in their new occupation. The peach and apple trees were ripped out because there was no market for their fruit. They were replaced by vines. Dried grapes were transportable and there was a market of sorts for them. Salting first appeared on some Vinifera blocks during the First World War. Some of the higher blocks were drained with tile drains soon after, but for most settlers, little was done about the problem. Other farms were on lower ground which was prone to salting and had clay soil which could not be drained. Farms should have never been located on such land. Elsewhere, tile drainage was quite possible, but settlers could not agree on whose property the main drain would be placed. The law said clearly that drainage works would only be built where there was 80 per cent local approval. It took 30 years for Nyah, the settlement founded on cooperative ideals, to agree on a drainage line.

By the mid-1920s, it was clear that soldier settlement was a financial disaster. In Nyah, it was also a land degradation disaster. During the Royal Commission into failed soldier settlement in 1925, both farmers and the Commission were keen to allocate the blame for the problem on salinity. Evidence was given to the Commission that in Nyah and Woorinen, a dozen soldiers had walked off their farms, many because of salt. The majority report blamed the failures in Nyah and Woorinen on lack of capital and lack of skill. The minority report, put out by the soldiers settlers' representative, pointed out that the 1916 enquiry into failed closer settlement policy had warned about the dangers of irrigating certain soil types, but that this had been ignored by the Commission. Worse still, blocks of land identified as unsuitable for closer settlement in 1914 were later sold to unsuspecting returned soldiers. The soldiers agitated for compensation for all settlers sold unsuitable land. This set the tone of relations between farmers and the State Rivers and Water Supply Commission (SRWSC) for the next 50 years.

In the years before the Second World War, Nyah struggled with salinity and marketing problems. Considerable areas of low-lying land were abandoned because of salt. Despite these losses, there was a consensus among the farmers and scientists that tile drainage would be economic and effective. Unfortunately, there was no consensus on the manner of implementation. Nyah farmers argued over where the drain should go. War years treated the dried fruit industry well. The price of dried fruit rose because of decreased production in war-torn Europe, just as it had during the First World War. Farm incomes rose by more than 60 per cent (McIntyre 1948 p 52-3). Many horticulturalists along the Murray Valley had a rare rest from worrying about making ends meet. This must have given them time to think about other problems (McIntyre 1948 p 44). More money may have helped the Nyah landholders finally come to some agreement about drainage. But by then, the job had to wait until the war finished and men and materials were available to do the work.

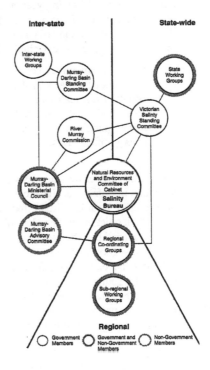

Figure 7.3   A system for coordinating a salinity control strategy

First Attempts to Sort Out the Problem

The Dried Vine Fruits Enquiry

While the end of the war promised some relief from Nyah's salinity problems, developments in the dried fruit market soon threatened the financially fragile settlers with ruin. Prices held up for a while, but the Swan Hill district had a series of wet, cold summers and autumns. The fruit did not dry, but rotted on the vines. Then, when the sun came out and warm summers returned, the price of dried fruit crashed, just as it had in 1923. The dried fruit settlements around Swan Hill could not compete with the warmer Sunraysia district, let alone the rest of the world. In 1950, the Australian Dried Fruits Association asked the federal government for help to restructure the settlements around Swan Hill. It took until 1954 for the government to respond. The Committee of Enquiry put out its report the following year, and it was not flattering in its assessment. The natural advantages of Nyah, over which Robertson had waxed lyrical in 1912, did not exist. It was too cold and wet by world standards. Even the Sunraysia district was regarded as marginal dried fruit country.

> While the committee considers that it is possible to carry out dried fruit production at a profit in the Mid-Murray area under certain conditions, it realises the activity is more hazardous than in other areas, both established and prospective. It therefore considers that further plantings in the Mid-Murray should be undertaken only on a very limited basis.
>
> *(Anon 1956)*.

The slightly cooler weather at Swan Hill meant the grapes ripened three weeks later than in Sunraysia, significantly increasing the chance of rain falling on a mature crop. Rain has little effect on immature grapes, but splits ripe grapes which then quickly rot. The Committee also noted the poor state of many of the blocks ravaged by salt. Despite these concerns, the Committee considered there was enough good land remaining to justify saving the settlements. But this could only be achieved by diversification to new enterprises and a massive restructuring of the settlements. There were too many small farms. It was generally accepted that a farmer needed 8–10 hectares to survive on an acceptable income. In Nyah, 79 per cent of blocks were smaller than eight hectares. Only 29 of the 161 blocks were judged to be viable, the others having too little land, too much debt or too much salt (BAE 1956 p 21).

The recommended solution was drastic. Many farmers needed to be persuaded to quit their farms for compensation of about $1400 per hectare, but any farmer refusing to take part in the scheme would not receive any other government assistance. The farmers that remained were to be encouraged to diversify their production, though into what was not exactly clear. Citrus, apples, pears, peaches and plums were unsuitable because of the risk of glut, lack of cannery facilities or inability to cope with the soil salinity. Prunes and dried apricots might work. Vegetables sounded better, but they needed a better water supply than the SRWSC was providing for the growers, namely water from the open channel systems on fixed rosters, with the time between waterings too great for the needs of vegetable seedlings.

The Committee of Enquiry report and its recommendations were not popular with the landholders. While half the farmers in the region were unhappy with their blocks, only 13 per cent actually wanted to sell up and move out. Some wanted to make a living on their blocks, rather than leave for an uncertain future as a labourer in the city. Others would sell up when the price of their property rose. In the meantime, the obvious thing to do was hang on in the short term, even if that meant making compromises about the long term. With money short and landholders again worrying about how the next bill would be paid, the problems of salt receded in the minds of many Nyah irrigators. For the government, it was politically easier to solve the salt problem than the block size problem. The wartime consensus on drainage had cleared the way for the SRWSC to build a drainage system to which farmers could connect their property drains. This was completed in 1949, but by 1956, many farmers had still not constructed drains on their own properties, despite obvious salinity and erosion problems (BAE 1956 p 22). They simply could not afford to do so.

Likewise, there was little rush to pull out the vines to grow alternative crops. In 1954, 90 per cent of Nyah farms had over 80 per cent of their plantings devoted to dried grape varieties. Today, Nyah shares its land between wine, sultana and table grapes. The result was unfortunate for Nyah, as the sultana has remained an unstable commodity. By the 1960s, nearly three-quarters of Australia's production had to be sold on the export market, but Australia's grapes made up less than 10 per cent of the total world trade in sultanas. This meant that Australian production had little influence on the world price. If there was a glut of production from Turkey or Greece, the price would fall dramatically. If that happened to be a bad year for Australia, or Nyah, then poverty on the farm was the inevitable result. Downstream, the Sunraysia and South Australian Riverland districts suffered similar hardships. Every time there was a price slump, there was another enquiry.

Putting Faith in New Products and a Piped Water Supply

In the 1960s, the Bureau of Agricultural Economics produced a series of reports on the poor prospects for the dried vine fruit industry. In 1966, it reported that severe income problems continued, especially in the Mid-Murray, where farm incomes were 30 per cent below sultana farm incomes in Sunraysia; Mid-Murray farms were half the value of sultana farms elsewhere along the Murray (BAE 1966). Industries Assistance Commission (IAC) enquiries in 1971 and 1976 considered similar issues. The problems remained essentially the same, namely inadequate block size and the wrong product. Again, it was easier to avoid the issue of block size and concentrate on the more palatable solution, a new agricultural industry.

Various approaches were tried to stimulate Nyah. Extension workers tried to promote new crops and improved growing practices. They examined how farmers in the district learnt their farming techniques (Presser 1972). In the 1970s, the Sun Centre Horticultural Committee tried a new marketing approach, selling the local produce under the *Sun Centre* label. In 1976, the Industries Assistance Commission concluded that 30–40 per cent of Sunraysia growers had little hope of becoming viable, commenting that the situation was worse in the Mid-Murray. It recommended quota protection and 'rural adjustment' assistance, including concessional loans and small grants to those leaving the industry.

The IAC had a number of criticisms which were particularly pertinent to Nyah. It criticised furrow irrigation because of the inherent inefficiencies of scale. An irrigator could only manage a 12 hectare farm irrigated with furrows. It favoured conversion to the new drip irrigation, supplying water to the trees through drippers in a plastic supply pipe – a simpler system, the operation of which did not involve channel formation and then waiting for flows to reach the bottom of runs. With spray or drip irrigation, one person could manage 40 hectares, a much more viable proposition given the prevailing commodity prices. The IAC also argued that furrow irrigation increased water accessions to the watertable and led to increased salinity. It also recommended that transferable water entitlements be introduced and supported pumped and pressurised water supplies.

The local Nyah Development Committee lobbied hard and successfully for a piped water supply. Because pipe supplies did not leak, water could be left in the pipes for prolonged periods between applications without any loss. This meant the SRWSC could supply water to a single grower at any time of the year without needing to refill the supply channels. Individual growers could then try to grow vegetable crops which needed watering outside the irrigation season. This would release Nyah from the strictures of a channel irrigation system. Nyah growers would be able to choose from a wide choice of crops instead of the low returns from dried fruit. The government accepted the pipeline proposal as the means of saving Nyah from financial depression. The SRWSC installed a large pump on the river and a piped water supply to every Nyah farm. In tones reminiscent of Robertson (1912), a market research report expressed optimism for the future of the Nyah District: 'Future years should see the area go ahead in leaps and bounds, particularly now the pipeline is on its way' (Presser 1972).

Paying Farmers Not to Farm: The Vine Pull Scheme

Many Nyah growers seemed uninterested in diversifying, however. When prices dropped, the logical step seemed to be to hang on and then sell out when sultana prices went into a temporary rise. The banks supported this strategy: this was the best way for them to get back their mortgage funds. The apparently low farm prices attracted buyers seduced by the dream of rural Utopia, and the cycle continued. In the next slump, the same properties were still not viable, but different people owned them. During the mid-1980s price slump, the IAC held another enquiry. A decade earlier it had pinned its faith on technology and new products. Now it decided more drastic action was needed. It proposed a 'vine pull' scheme. Growers would be paid money to pull out their vines and sign a contract not to replant vines in the same paddock for another six years. The theory was that the farms would then be planted to some other crop. Again, the real world did not behave according to theory.

The Vine Pull Scheme has reshaped Nyah, but not in the manner anticipated. To some landholders, the Vine Pull Scheme offered an opportunity to be paid to rip out old vines which should have been pulled out years ago. To others, it was an opportunity to get quick cash. The vines were pulled and the land sold. Now this land remains vacant. No new crop grows on the land. The owners of the land still have to pay their water bills. Some growers who ripped out their vines planted peaches and nectarines. Production naturally increased and now peaches and nectarine prices are as low. Many vegetable prices are not much better. Following the merger of major wine companies, wine grape prices have also dropped.

As this study is being written, the horticultural and vegetable industries are suffering their lowest real returns in 40 years. The small farm problem moves from one commodity to another. Some grower organisations blame the overproduction on 'refugees' from the city rat-race, lured by dreams of Utopian rural lifestyles. They have established or rebuilt small farms and increased production beyond the demands of the local markets.

SECTION 2: WHAT IS TO BE MANAGED IN THE NYAH DISTRICT?

Agriculture

Dryland Farming

The Nyah Irrigation District is surrounded by mainly sandy, easily erodible, undulating Mallee country, which has largely been cleared for dryland farming. The area's average annual rainfall is approximately 300 mm. The main crops grown are wheat, barley, field peas, lupins and medic-based pastures. The livestock industries are primarily prime lamb and wool production and some beef cattle.

Farming in the Nyah Irrigation District

The Nyah Irrigation District covers approximately 1560 ha, just over 1000 ha of which are irrigated. About 750 ha are planted to either annual or perennial horticultural crops and 250 ha under irrigated pasture. The remainder is used mainly as dryland mixed farming land. The main horticulture crops are wine grapes and vegetables, followed by stonefruit and citrus. Small areas of other crops are also grown (Table 7.2).

Table 7.2   Horticultural crops in the Nyah Irrigation District 1987

Vine and tree crops	Area (ha)	Vegetables	Area (ha)
Grapes	517.2	Pumpkins	19.2
Dual & Multipurpose	481.4	Melons	15.0
Table Grape	35.4	Rockmelon	7.6
Stonefruits	74.9	Watermelon	2.8
Apricots	36.6	Honeydew	3.8
Nectarines	13.8	Other melon	0.8
Peaches	8.3	Zucchinis	2.8
Plums	16.2	Cucumber	1.2
Citrus	19.4	Squash	2.2
Olives	3.8	Tomatoes	3.2
Persimmons	1.0	Broccoli	14.0
Cherry	1.1	Cabbage	2.6
Nashi pears	0.7	Onion	2.4
Apples	0.7	Garlic	4.0
Figs	0.8	Lettuce	0.4
Walnut	0.7	Other	7.6
Almonds	1.7	Seed Veg:	15.0
		Lettuce	4.4
		Onions	8.4
		Cauliflower	2.2
Total	**662.0**	**Total**	**89.8**

Source: SCHDA 1987

In addition to the horticultural crops, a significant area of the Nyah Shire is under pasture. Of this, approximately 250 ha are irrigated perennial, annual and lucerne pastures. There are only a few full-time graziers. Most of the farms growing pastures are hobby or part-time farms and mainly run cattle and some sheep and horses. Most of the cattle herds are small, being less than 30 head. Some of the pastures are used for baled hay, both for on-farm use and for sale.

In addition to the irrigated pasture land, the district has about 540 ha of dryland grazing land, reserves and unused land. This large area of unirrigated land is mainly due to the size of irrigation entitlements, the landholders not having enough water to irrigate all their land.

Farm Size in Nyah Shire

Table 7.3   Number and average size (hectares) of farms in the Nyah Irrigation District

Farm type	Number	Average size (ha)
Mixed	72	6.0
Horticultural	129	8.9

Source: Swan Hill District Horticultural Census 1987; Rural Water Commission Culture Sheets 1989/90

Table 7.4   Farm size of horticultural properties in the Nyah Irrigation District 1987

Farm size (ha)	Number of category	Percentage in each	
0–5	26	20	
5–10	66	51	
			71
10–15	20	16	
15–20	12	9	
			25
20–25	2	2	
25 +	3	2	
			4
Total	129	**100**	**100**

Source: Swan Hill District Horticultural Census 1987

The 1987 Swan Hill District Horticultural Census identified 129 horticultural properties in Nyah Shire irrigation area, with an average size of 8.9 ha (see Table 7.3). Table 7.4 presents the frequency of farms by area.

In addition to the horticultural properties, there were 72 'mixed' properties with an average size of 6.0 ha. These are properties where the main land use is not horticulture, though some horticultural activity may be present.

Irrigation Practices and Management

The areas irrigated by differing systems of irrigation are given in Table 7.5.

Table 7.5   Irrigation systems, by area, in the Nyah Irrigation District 1990

Irrigation system	Area (ha)	Percentage of total (%)
Flood	228.0	21
Furrow	747.4	70
Sprinkler	82.3	2
Moving	6.5	<1
Overtree	74.7	7
Undertree	1.0	
Microdrip	17.1	2
Total	**1074.8**	**100**

Source: Rural Water Commission Culture Sheets 1989/90

In the Nyah Irrigation District, flood irrigation is used for irrigating annual, perennial and lucerne pastures. Some lucerne pastures are also irrigated by moveable sprinkler systems. Most of the horticultural crops, including nearly all grapevines, are also irrigated by furrow irrigation. Overhead sprinklers are used to irrigate citrus, some vegetables and some vines. Moveable sprinklers are mainly used on irrigated lucerne, but also on some vegetable crops. Microdrip irrigation is used on stonefruits and vegetables.

Efficient irrigation involves applying only as much water as the plant can take from the soil and at the right time, ie before the plant becomes stressed through lack of water. A number of techniques can be used to aid irrigation scheduling in horticultural crops. The main ones are tensiometers, test wells and evaporation information. However, it would appear that in Nyah, most growers tend to rely on their 'gut feeling' rather than any of the methods mentioned. Tensiometers are simple and reliable tools which warn a grower before plants become stressed and are one of the best available tools for scheduling irrigations in furrow irrigated orchards. However, only a small number of farms use them. Also, they are often used incorrectly, or even if installed, not used at all. Test wells, holes in the ground which are used to measure the depth to the water table, are used to identify problems with over-irrigation and poor drainage, but they are rarely used in the Nyah District. Evaporation information, together with the crop factors, is used to calculate how much water a crop has used over any given period, and thus when the next irrigation should occur and how much water should be applied. Evaporation information was formerly supplied each week in the *Swan Hill Guardian*, but was little used.

As a general observation, grape and stonefruit growers tend to over-irrigate in spring and under-irrigate during summer. This is partly due to 'tradition', a lack of information on crop water requirement during the season, and the slow adoption of irrigation scheduling methods which would more accurately determine irrigation needs.

Traditionally, irrigation has been started around August/September. Some irrigation is applied during the winter for frost protection. However, there is a compromise between frost protection and a healthy root environment. The tendency to over-irrigate early in the season compounds other problems such as lime-induced chlorosis and salinity. Underwatering of crops is common during summer, especially with sprinkler and drip irrigation systems. Drip irrigation systems require frequent water applications. The most common problems with furrow irrigation management are associated with poor irrigation layout, and when to start and stop irrigation. Another problem is that growers may also need to wait for their turn to get water during peak demand.

Agricultural Productivity

Crop productivity is determined by a large number of factors and their interaction. With particular reference to horticultural crops, these include varieties, soil and irrigation management, pests and diseases, availability of cool stores and efficient grades. Little specific information is available on such factors in the Nyah District.

Agricultural productivity varies widely between properties in the Nyah Irrigation District. Overall, however, productivity is low. This can be accounted for by such factors as small farm size, high turnover of farm operators, and the large percentage of hobby and part-time farms. According to Chamberlain (1988), many of the new arrivals have little or no farming experience and little money to put into farm development. This results in the district having little

119

family farming tradition and, with the lack of community leaders or early adopters, little farming innovation. The high number of hobby farms also makes change difficult, as does the small farm size in generating profits for development.

Market Practices and Structure

- **Viticulture:** Grapes are marketed as fresh table grapes, winegrapes, dried vine fruits or canning grapes. The majority of the table grapes are marketed domestically, with the wholesale markets in Melbourne, Sydney and Brisbane being the main outlets. Commonly, part of the crop is stored for later sales in coolrooms on growers' properties: there were 23 such coolrooms in 1987. Winegrapes are sold to wineries in various locations, including Swan Hill, Sunraysia and the Murrumbidgee Irrigation Area. The dried vine fruits from the Nyah area are marketed mainly through the Mildura Producers Cooperative and, to a lesser extent, the Robinvale Producers Cooperative. Some grapes are sold each year for canning purposes to companies in the Shepparton area and Leeton.
- **Stonefruits:** As with table grapes, most stonefruits are marketed domestically on the wholesale markets in Melbourne, Sydney and Brisbane and to lesser extent in Newcastle and Tasmania. A large quantity of stonefruit is exported to South East Asia, the Pacific Islands and the Middle East. Dried stonefruit are marketed through packing companies in South Australia.
- **Citrus:** Most of the district's citrus is sold through the local Marangan Packers Pty Ltd, which packs and sells fruit on the local and export markets, as well as juicing fruit in its own plant. The company is a registered exporter and sends citrus to Europe and Asia. However, most of the exported fruit is grown outside the Nyah District.
- **Vegetables:** Most of the area's vegetables are sold on the domestic markets, primarily through the capital city wholesale markets. There is also some direct selling by growers, both to retail outlets and through roadside stalls.
- **Vegetable seeds:** including onions and lettuce, are grown under contract to seed companies.
- **Pasture, hay and livestock:** Some pastures, mainly lucerne, is baled and sold either in the local area or elsewhere. The majority of pasture is used for grazing beef cattle, sheep and horses. The cattle and sheep are sold through the Swan Hill saleyards.

Secondary Industries

While the Nyah Irrigation District has a number of support and service industries, there are only two establishments that can be considered secondary industries. One is Marangan Packers Pty Ltd, which, as noted above, in addition to packing fruit, also produces citrus juices. The other is Lloyd's Vineyards, which make organic grape juice and other dried organic products.

Service and Commercial Industries

The agricultural and horticultural interests are well supported by general engineering and hardware establishments. The building and associated trades are also well represented as is the automotive trade. In addition, many tradesmen live and work in the district. There is also a range of commercial activities. In addition, the district has other major facilities available in nearby Swan Hill.

Community Services

The small townships in the Nyah area are typical of hundreds of small rural towns across Australia. A wide range of community services are testimony to the efforts of citizens to help themselves. However, many of these services are in decline and the towns show signs that they are not as vibrant as they were even ten years ago. Nevertheless, a community spirit still exists in the area. People are prepared to change their ways to ensure the survival and development of the district.

There are three primary schools which cater for approximately 180 pupils. The district has numerous activities for pre-school children and various youth groups, though the latter are not being utilised to their full potential.

The District has a hospital, the Nyah District Bush Nursing Hospital, a Day Care Centre and a doctor who resides in the area. The Nyah West and District Senior Citizens are a very active group who provide a wide range of functions to entertain people during their retirement years. Meals on Wheels also provide a service for the elderly.

There are numerous sporting venues in the area. A major trotting complex is situated at Nyah, where football and cricket are also played. At Nyah West there is a sports oval. There are two sets of tennis courts, two golf courses, lawn bowls, a gliding club and a pistol club. Netball is also played and at Irrigatia there is a pony club.

The Nyah District has much to offer the tourist, for example the Nyah and Irrigatia State Forests, the old steam pump on the Murray River and the horticultural setting. There is accommodation at the caravan park at Nyah and the Grand Hotel at Nyah West.

SECTION 3: HOW IS NYAH SHIRE BEING MANAGED?

Shire Planning Controls

The Council of the Shire of Swan Hill (the Shire) is the responsible authority for the administration and enforcement of planning controls in the Nyah Irrigation District. No planning control existed prior to 1964. The Shire of Swan Hill Interim Development Order (IDO) Planning Scheme was gazetted in 1964 and covered the townships of Nyah and Nyah West and the horticultural and dry land farming areas that existed between and around the two townships.

The IDO Planning Scheme limited subdivision for the rural areas into allotments of not less than two ha, while subdivision minima around the remaining area (townships) was not spelt out, but was at the discretion of Council. In 1982, the Council refined its planning controls for rural areas, by adopting a policy for the horticultural areas of the Shire. A landowner was able to apply to Council for a planning permit to subdivide a lot of less than 9.7 ha off his/her horticultural property as long as the balance of his/her property was greater than 9.7 ha. The objective of this policy was to allow the landowner to subdivide a house block off and sell the balance as a viable horticultural property. The basis for selecting an area of 9.7 ha as being a viable property for horticultural purposes was based on government advice at the time.

In 1987, Council modified this policy, to set a minimum lot size for subdivision in the horticultural areas to 14 ha, with no small lot subdivision being permitted. However, Council only maintained this new policy until 1988, when it returned to a variation of the earlier horticultural policy. Under this further modified policy, the small 'house' lot excision (previously less than 9.7 hectares and later prohibited) was now set at between 0.2 ha and 0.6 ha, provided that the balance of the subject land was still greater than 9.7 ha. This policy is still in operation.

There has been little subdivisional activity in the Nyah Irrigation District, possibly due to the fact that most horticultural properties are less than 9.7 hectares, or that larger rather than smaller properties are often sought after for viable agricultural pursuits. Planning appeals against subdivision inconsistent with Council's controls and policies have generally been disallowed.

Irrigation Management

Under the Water Act 1989, the Rural Water Commission (RWC) is responsible for the following functions for each of the constituted Irrigation Districts:

1 To provide, manage and operate systems for the supply of water to irrigable lands and for the appropriate drainage and protection of those lands;
2 To identify community needs relating to irrigation, drainage and salinity mitigation, and to plan for the future needs of the community relating to irrigation, drainage and salinity mitigation;
3 To develop and implement programmes for improved irrigation practices, improved drainage practices and improved salinity mitigation practices;
4 To investigate and research any matter related to its functions, powers and duties in relation to irrigation, drainage and salinity mitigation.

The Nyah Irrigation District consists of 1566 ha of which 1096 ha is deemed commanded (land that can be gravity supplied from the pipeline system) and suitable for irrigation. Water rights have been allocated on the basis of 7.62 ML per ha of commanded and suitable lands. As at 1 July 1990, the total water rights allocated amounted to 9411.1 ML, plus 248.8 ML for stock and domestic purposes.

The Commission's assets within the Nyah Irrigation District include land, pipelines, drains and pump stations, with an estimated written down current value of some $8 million.

Water is pumped from the Murray River by the Nyah Main Station, located on the banks of the River. The station consists of five 25 ML/day pumps with a combined capacity of 125 ML per day. To provide for some variability in demand, two pumps are variable speed types.

The pipeline reticulation system was basically installed along a similar alignment to the original open channel system and consists of 51.2 km of concrete pipelines varying in diameter from 1200 m to 255 mm.

Irrigated farm lands are generally tile drained with the effluent being directed to the Commission's drainage system. Basically the subsurface drains service seven separate catchments which generally outfall to the forest areas adjacent to the Murray River. In addition, there are some 250 ha which drain naturally to the forest area.

During the winter months, floodwaters inundate the forest area and flush the lagoon systems, as well as leaching accumulated salts from the soils. It is estimated that in a normal year, some 6670 ML are pumped into the Nyah Irrigation District at an average salinity of 240 EC or 960 tonnes of salt per annum and that 800 ML having an average salinity of 1600 EC or 770 tonnes is returned to the river each year.

The Commission is required under the Water Act to make and levy irrigation charges each year to provide for:

1   liquidation of losses incurred in the previous three years;
2   operation and maintenance costs for the water supply and drainage system;
3   replacement and enhancement of assets.

The annual water rates in the Nyah Irrigation District have increased from $231 000 in 1981/82 to $404 000 in 1989/90. Throughout this period the proportion of the rate that has been collected has been approximately 85 per cent.

Conservation and the Environment

Within the Shire are various areas of public lands, the overall responsibility for which rests with the Victorian Department of Conservation and Environment. These lands include areas of state forest, the River Murray Reserve, and recreation reserves.

The Nyah and Vinifera State Forests, located along the Murray River, include stands of river red gum, black box, mallee and cypress pine. The forests support grazing and timber production, both of which are closely monitored to prevent degradation of the natural environment. The forests also include over 170 aboriginal sites, most of them oven mounds. The River Murray Reserve includes a 60 metre wide Public Purposes Reserve along the River. Within the District, the Nyah Golf Course is classified as a Recreation Reserve.

The Community's Decision to Change

With the preceding information documented and the implications of the future looking bleak, the Department of Agriculture and the Rural Water Commission jointly called a public meeting chaired by the local State Member of Parliament in July 1989. The meeting was particularly vocal about water charges. By the end of the evening, however, after the issues had been discussed, it was decided to form a community group. The members of the group were nominated from the meeting. The group was named the Nyah District Action Group. The Group was made up mainly of horticulturalists from various sized horticultural blocks.

When the steering committee of the Nyah District Action Group was first formed, a submission was made for funding under the Natural Resources Management Strategy of the Murray-Darling Basin Commission (MDBC). As part of the strategy, funds are made available for various kinds of projects throughout the Basin. In the Nyah case, the submission was made under Category 10A: *Water and Land Management! Establishment of Communities of Common Concern.* The project title is 'Salinity Management, Rural Renewal Strategy Plan, Nyah Irrigation District'. The objective of the project is 'To develop a sub-regional management plan that provides for no net increase in drainage disposal to the Murray River while markedly increasing horticultural production and productivity'. In other words, the group's objective is to develop a Salinity Management/Rural Renewal Plan for the Nyah Irrigation District.

The Nyah District Action Group began to meet regularly, with Neville Harris as elected chairperson. In April 1990, a Community Development Officer was appointed to assist the group. To meet the MDBC guidelines for the development of such plans, there was a requirement for a broader community representation. The Action Group then saw itself as a steering committee until the new committee was elected. During this period, the Community Development Officer conducted a survey of the needs of the community and its views and thoughts for the future.

Another public meeting was held in October, 1990, to elect a working group. The representation of the new Nyah District Action Group is six horticulturalists, two community representatives, two urban dwellers, two business people, a Shire of Swan Hill representative and a person representing environmental issues. The meeting was not well supported, but there was enough interest to fill the positions and form and endorse the working group.

The working group developed five basic goals for the future of the Nyah Irrigation District.

1   To improve the image of Nyah to people outside the area.
2   To improve productivity of horticultural crops.
3   To increase the availability of year round employment.
4   To increase services available in the community.
5   To maintain current services.

As the Action Group progressed, it developed a mission statement and strategies taking the objectives and the goals into consideration.

The mission statement reads:

> 'By public involvement to develop sustainable horticultural industries, while reducing salinity inflows to the Murray River, to improve the status of the Nyah Irrigation District Community.'

The following seven strategies were developed:

1   Improve productivity of horticultural crops and introduce new crops;
2   Encourage horticultural practices aimed at preventing degradation;
3   Improve employment opportunities;
4   To improve the perceived status of the Nyah Irrigation District;
5   To maintain and increase services within the community;
6   To encourage efficient water usage practices aimed at reducing saline discharges to the Murray River;
7   To create an awareness and understanding of environment issues within the local community.

The Nyah District Action Group formed four sub-committees to investigate the various problems in the district in the light of the goals and strategies. Because there were seven strategies, however, the Group decided to combine Strategies 1 and 2 with one sub-committee looking at cash and established crops, herbs and land and water management. Strategies 3, 4 and 5 were combined with a newsletter sub-committee and another sub-committee of social, light industry and tourism. Strategies 6 and 7 have a subcommittee investigating environmental issues.

These sub-committees developed an options paper (set out below) which was put before the Nyah District Action Group and tabled at the public meeting on December 1991. The information gathered from a survey of every household in the Nyah District was also to be presented at this meeting. In this way, the community believed it was scaling the 'Conflict Management Mountain' to the point where the public meeting would provide opportunity for coordination and mediation.

SECTION 4: OPTIONS FOR THE NYAH DISTRICT

'Do nothing' Scenario

Advantages

- The district would continue to slowly become a dormitory community of Swan Hill.
- If less water is supplied to the District due to the change in cultural practices, then less water would return to the river as drainage. This would lessen any environmental degradation that may occur over time.

Disadvantages

- If there is no intervention, the Nyah Irrigation District will continue to decline. It will remain an area of large, commercially unviable hobby farms (too big to mow, too small to grow). The land would be lost to horticultural production.
- There is little financial incentive to encourage restructuring by amalgamation. This could lead to a loss of services, both to the remaining horticulturists and also to the community in general. These would include government services, the hospital and doctor, service industries, and retail businesses.
- The population is static if not declining. At present, approximately 11 per cent of the land changes hands annually, limiting the ability to build a core of experiences producers. This turnover of property owners will continue.
- Generally, returns from the sale of land would not be enough to purchase a house in one of the townships. For this reason, many horticulturists who may wish to retire, remain on their blocks and cut back on production. The result is an ageing population and declining agricultural output.
- The Rural Water Commission has an expensive system to supply irrigation water to the district. With continued reduction in horticultural production, the economic return on the capital investment would be further eroded. The financial burden of continuing the operation of the irrigation system would have to be carried by the landholders even though the value of the grazing would not cover the cost of the water.

Encourage the Development of Hobby Farms

Advantages

- Hobby farms would result in an increase in population and in rate revenue to local government. Further subdivision also has the potential to increase land values. It would also increase water rate revenue, because the Rural Water Commission could increase the rate to that for Domestic and Stock (Rural Residential). This would justify the continuation of the system.
- There would also be the potential for increased spending within the district. Professional people would be attracted to the area to buy subdivided blocks for a rural lifestyle. They would be financially secure, generating a regular income. They may attract other service industries.

Disadvantages

- There would be an accelerated loss of horticultural land to subdivision. There may also be the loss of some of the horticultural services. On the other hand, some capital intensive farms may not recoup their money.
- There would also need to be strict policy on dwelling requirements for the subdivided blocks to remove the fear of 'shanty town' development. The high cost of developing the infrastructure to service the subdivisions (eg access, power, water, etc) could be a problem, as could waste disposal.

Coexistence of Commercial and Hobby Farms

Amalgamation and Subdivision within Existing Area

Advantages

There are numerous advantages for amalgamation and subdivision:

- It would overcome the problem of the intermediate block size. The majority of blocks are of this size and are causing most of the problems in the District. Many people buy blocks of four to eight ha with great expectations, only to be caught with a property that is too small to divide and too small to be economically viable. If subdivision was freed up and assistance for amalgamation provided, many of the expectations could be met. In particular, a horticulturist could expand his holding and/or subdivide his house from his block, sell the block and retire in the house.
- A major advantage would be that it would reduce the turnover of farms in the district. It would ease the financial difficulties associated with purchasing a block to increase farm size. It also has the potential to increase production as the horticultural land would be farmed more efficiently by professional farmers.
- Rate revenue would be higher, due to hobby farms and a higher standard of commercial farms. There would be a greater use of the available water and a greater percentage of water rates paid. The increase in general value of production would at least maintain services and employment, casual, part-time and full-time. The community would have a social feeling of well-being and economic security. Better management practices, horticulturally, capital maintenance, weed control, fire hazard, and so on, would occur.

Disadvantages

- Infrastructure costs to accommodate this transition could include water connection etc.
- In future years, conflict may arise because of the horticultural practices of commercial farming and the rural retreat philosophies of others.

Zoning Rural Residential in Future

Advantages

The advantages of zoning rural residential are that the poorer soils and frost prone areas could be targeted. Well defined areas could be marketed for rural residential.

This would minimise conflict between rural residents and commercial farmers. Instead of scattered development of hobby farms, the economies of scale for infrastructure would be far more attractive. As indicated with other options, it would increase diversity within the district, contribute to a younger population, increase local government rates, and help maintain existing services.

Disadvantages

Inequity in land values would result from the creation of a designated or zoned area. Pressure could be applied to service such new rural residential areas at the expense of the existing ones.

In particular, this does not address the structural adjustment problems outside the designated area.

Expanding Hobby/Rural Residential outside the Defined Irrigation District

Advantages

As above: areas which have poor soils or are frost prone could be marketed for rural residential, thus increasing diversity in the district as well as wealth and services.

Disadvantages

There is no existing infrastructure and so the costs of development would be high. By establishing a new area outside the Nyah district, services within the district may not be properly utilised. Later on the new area may become an isolated ghetto. There is the possibility of environmental degradation.

Expanding Horticultural Farming outside the Nyah Irrigation District

Advantages

- This would eliminate consolidation difficulties for new farmers. There would be no existing management problems to overcome. The larger horticultural blocks would encourage entrepreneurs to the district. This would add to the nucleus of professional farmers which would increase both production and productivity. There could be the opportunity for lower unit water costs with the increased quantity of water supplied. Hobby farmers would be excluded from the new development, so eliminating this potential source of conflict.
- There would be increased employment opportunities for the existing population of the Nyah district. There would also be increased spending power within and outside the District and an increase in local government rate revenue. There would also be flow-on effects of both values and confidence to the existing District.

Disadvantages

- The increased demand for water could reduce its availability during peak demand periods. Provisions would need made to dispose of the increased quantities of drainage water. What would be done with this? There may well also be effects on groundwater levels.
- There would be a need to extend the existing infrastructure.

Increase Commercial Farming and Reduce Hobby Farms

Advantages

These are mainly of an economic nature, including an increase in production (and hence its gross value) and the optimum use of the Rural Water Commission infrastructure.

Disadvantages

Because of the existing makeup of farms, hobby, part-time and commercial, it would be almost impossible to undertake such a change. As a consequence, there would be a population decline which would reduce the labour supply and there would almost certainly be a decline in existing services.

SECTION 5: WHO IS TO MANAGE THE ENVIRONMENT IN THE NYAH DISTRICT?

The integrated environmental management necessary to bring the Nyah District to a sustainable status requires and will continue to require considerable individual and organisational learning and cooperation. The Nyah District management team is not only dealing with economic and environmental change in the District, but also with social and individual change. Representatives of the five groups necessary to achieve sustainable economic, environmental and social change have been involved in the Nyah District renewal planning. Examples of these representatives are outlined in Table 7.6.

Table 7.6 Sustaining economic, environmental and social change in the Nyah District – the five key groups

	Policy-makers	
Nyah District Action Group	Sue	horticulturist
	Gavin	horticulturist/vegetables
	Ian	horticulturist
	Anne	teacher and hobby farmer
	Tim	horticulturist/seed production
	Bill	Shire Councillor and Mallee farmer
	Ian	teacher and hobby farmer
	Ross	chemist and hobby farmer
	David	Shire of Swan Hill Economic Development Officer
	Brian	horticulturalist/Australian Dried Fruits Association representative and member of the Irrigators Advisory Group
	Collin	horticulturist
	Neville	horticulturist and owner of a Packing Company
	Malcolm	Nyah town resident and horticulturist at Woorinen
	Ian	retired resident
	Ken	Community Project Officer
Technical committee	Ian	Manager, Department of Agriculture (Swan Hill)
	Mark	Department of Agriculture (Mildura)
	Rob	Department of Conservation and Environment (Swan Hill)
	Les	Rural Water Commission (Swan Hill)

131

Greg	Rural Water Commission (Swan Hill)
John	Planning Officer, Shire of Swan Hill
Sue	Chair, Nyah District Action Group

Others outside the Nyah District

The Murray-Darling Basin Commission
The Victorian State Natural Resource
Management Assessment Panel
The Victorian State Salinity Bureau
The Victorian Farmers' Federation

Specialists and/or researchers

Bruce	Community Participation Consultant
Neil	State Department of Agriculture (Melbourne)
David	Economic Development Officer (Swan Hill Shire) Hydrologist
Extension Officers	Department of Agriculture, (Swan Hill)

Community members

Service Clubs such as Apex and Lions
Participants in the Public Meetings and Survey

Enterprise initiators

Bruce	Community Participation Consultant
Neville	Owner and Manager of the Packing Company
Ken	Community Project Officer
Tim	Horticulturist developing new horticulture skills Extension Officers

Service providers

| Government departments: | Rural Water Commission Agriculture Conservation and Environment |

Country Women's Association
Banks

References and Further Reading

Anon (1956) *Report of the Committee of Enquiry into the Mid Murray Dried Vine Fruits Area*

BAE (Bureau of Agricultural Economics) (1956) *Dried Vine Fruit Industry: Management Practices in Victoria and New South Wales* Bureau of Agricultural Economics, Canberra

BAE (Bureau of Agricultural Economics) (1966) *The Australian Dried Vine Fruits Industry: An Economic Survey* Bureau of Agricultural Economics, Canberra

Barr, N and Cary, J (1992) *Greening a Brown Land: the Australian search for sustainable landuse* McMillan, Melbourne

Chamberlain, N (1988) *Factors Affecting Some Farming Decisions in the Districts of Woorinen, Nyah, Tresco and Tyntynder* Report to the Rural Water Commission of Victoria, Melbourne

Government of Victoria (1987) *Salt action: joint action* Strategy for managing the salinity of land and water resources, Government Printer, Melbourne

McIntyre, A J (1948) *Sunraysia: a social survey of a dried fruits area* Melbourne University Press, Melbourne

Nyah District Action Group (1991) *Nyah Irrigation District Status Report, September, 1991* Department of Agriculture, Swan Hill

Presser, H A (1972) *Report to the Nyah Community* School of Agriculture and Forestry, University of Melbourne

Robertson, J E (1912) *The Progress of Swan Hill and District: introducing Ultima, Lake Boga, Nyah, etc* Melbourne

Rural Water Commission Culture Sheets 1989/90

SCHDA (Sun Centre Horticultural Development Association) (1989) *Swan Hill District Horticultural Census 1987* Sun Centre Horticultural Development Association, Swan Hill

Scholes, L (1989) *A History of the Swan Hill Shire: public land, private profit and settlement* Shire of Swan Hill, Swan Hill

Swan Hill District Horticultural Census 1987

SWRWSC (State Rivers and Water Supply Commission) (1983) *Nyah Irrigation District: operating procedures* State Rivers and Water Supply Commission, Swan Hill

133

8 Trouble with Traffic

The effects of rapid urbanisation on Bangkok, Thailand*

Anuchat Poungsomlee, Helen Ross and Rob Wiseman

Background

Bangkok is the capital of the South-East Asian country Thailand. It is home to 7–8 million people, and is the seat of the country's government and national administration; the bulk of the country's industry and trade; and virtually all of Thailand's telecommunications, higher education, public welfare and social services.

The Bangkok of the 1990s is largely a result of an aggressive policy of modernisation pursued by the country's rulers since the mid-nineteenth century.

Bangkok dominates urban life in Thailand. It is fifty times larger than the country's second largest city, Nakorn Rachasima. About 10 per cent of all Thais live in Bangkok. There are staggering economic and social differences between urbanised and rural Thais. The city is beset by enormous social, economic and environmental problems, including displacement of traditional landowners, traffic problems, air pollution, sprawling slums, and poor water quality.

---

* This chapter is based largely on the documents by Poungsomlee and Ross. The authors thank Sureeporn Punpuing and Suwattana Thadaniti for assistance with details.

This chapter explores one of the city's most visible and infamous problems – its transport system. The first section looks at the causes of the problem, and some of the consequences of the city's transport woes. The second section looks at the various organisations and individuals in Thai society that have a stake in maintaining and reforming the transport system. The third section examines some of the conflicts involved with the system and in changing it, and the fourth examines some of the ways in which changes are being made.

The Causes of Bangkok's Transport Problems

Geography

Bangkok is built on the delta of the Chao Phraya River. The floodplain is very flat and very low, with the entire metropolitan area only about one and a half metres above mean sea level. As a result, the city has a major drainage problem, with several parts of the city being submerged during the rainy season. In some areas, a heavy fall of rain and a high tide are sufficient to cause local flooding at any time of year.

The area is drained by a system of *klongs* (canals), dug in the first century of the city's life to add to the natural waterways of the delta. Thai society was formerly oriented around such waterways, which provided domestic and agricultural water supplies, waste disposal, and the main means of transportation. Two-storey houses, designed to cope with floods, faced directly onto the water. This style of housing remains prevalent in parts of Bangkok, where residents still move around their neighbourhoods by small boat and shop at stores which face onto the water.

The drainage function of the klongs is one of the casualties of Bangkok's form of modernisation. Many of the former canals have been filled in to provide road space, or have been blocked by unregulated development. At the same time, the remaining canals have to cope with vastly more urban run-off, as so much of the land surface has been built over.

Bangkok's drainage problems are compounded by the soil structure on which the city stands. Bangkok is built on an alluvial plain consisting of alternate layers of clay and sand-gravel. During the rainy season, the clay is heavy and muddy, but hardens during the dry season. The city's major structures are built on the topmost layer of clay. Unfortunately, this clay layer is very sensitive to vibrations, such as those caused by heavy traffic loads. Roads, buildings and other structures can and do crack within a short time under the constant vibration caused by road traffic. This problem has long been overlooked in planning, and the result is massive (although barely visible) damage to many public and private properties.

135

Underground water is located at depths of 150 to 200 metres. Due to the shortage of water in the city, a large amount of water has been pumped out. The result has been severe subsidence in many areas, particularly in central and eastern Bangkok. The heavy use of groundwater continues because of the high cost of providing an integrated water supply system over a large area, combined with rapid urbanisation in areas without existing water supply systems.

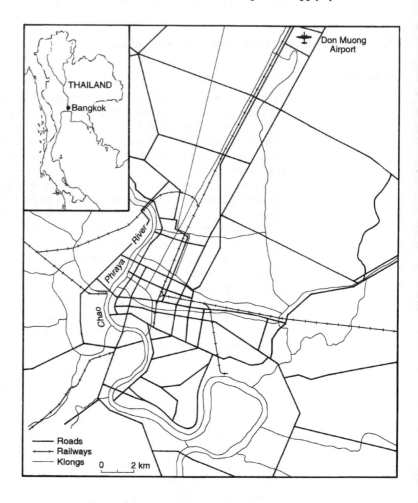

Figure 8.1   Bangkok – roads, railways and *klongs*

The administrative unit known as the Bangkok Metropolitan Area has never coincided with the built-up area of the city. The Metropolitan Area includes a high proportion of agricultural land, which remained even quite close to the city centre until the late 1980s. As well as being productive and enabling rural people to continue their ways of life, this land contributes significantly to the ecology of the city system. It contributes to the drainage system, by compensating for the built-up areas which cannot absorb water. The vegetated areas provide some compensation for, and relief from, air pollution. The mixed agriculture also provided a natural system of pest control, since neighbouring farms provided predators for one another's pests.

There has been gradual change in the produce grown, with rice predominating last century, then being replaced with a variety of fruit orchards and spices. These in turn are giving way for orchid farms, which can be air-freighted worldwide from Bangkok's international airport. Urban water pollution transported down the klongs is affecting food production seriously, but poses no threat to flowers.

The agricultural areas also have an aesthetic function, giving residents virtually the only access to greenery in a city with few parks. They are protected by planning measures, but are nevertheless being encroached upon rapidly.

A Brief History

Bangkok was established as the capital of Thailand in 1782, when it was known as Siam. At the time, most of the fertile floodplain around the site of modern-day Bangkok was devoted to agriculture – mostly rice, with some orchards. In the middle of the eighteenth century, rice amounted to about 95 per cent of national agricultural production (Ingram 1955).

The original city of Bangkok was very small – only 3.5 square kilometres. At the centre of the city was the palace and a complex of monasteries and temples surrounded by three concentric klongs. In the outer section were residential areas interspersed with paddy fields. Many residents lived in boats moored several deep along the river, and in traditional houses facing the klongs.

For the next century, the city grew comparatively slowly. The system of klongs used for transport in the city was expanded, but remained adequate for the city's transportation needs, as well as to serve the city with water.

In 1855, King Rama IV signed the Bowring Treaty with Great Britain and embraced modernisation, opening the country to the western world. The treaty was originally established to improve trade between Siam and Britain. The Thais, however, also had an eye on using similar treaties with other trading powers to avoid colonial exploitation as had occurred in other parts of Asia.

Skilled diplomacy with Britain, France, Denmark, the United States of America, Germany, Japan, Spain, Sweden and Russia and strategic efforts to modernise enabled Siam to remain the only uncolonised country in the region. Now, 150 years later, modernisation and westernisation are largely interchangeable expressions in Thailand.

As the Thai economy became integrated within the world economy through international trading relations, Thailand underwent increasing commercialisation. Initially, the country's major export was rice. However, this was increasingly supplemented by exports of natural resources and raw materials such as teak, tin and rubber. At the same time however, many home market industries, such as textiles, declined because they were unable to compete with cheaper imported products (Ingram 1955). The net result for Siam was a greater dependence on international trade and monetary exchange values. Despite its political independence, Siam found itself encumbered with a *de facto* colony economy – export of cheap raw materials and natural resources, with imports of luxury consumer goods rather than basic essentials. Nonetheless, the policy and process of modernisation continued apace.

The process of modernisation was centred almost exclusively on Bangkok, where industrialisation and commercialisation were increasingly concentrated. Meanwhile, people in the rural sector still pursued essentially traditional agricultural activities, and spoke their own dialects. The difference in paths of development taken by both groups also involved increased difference in social structures and ways of life. By the first decades of the twentieth century, there were marked differences between urban and rural Thai societies.

As Bangkok expanded, it swallowed up surrounding villages and agricultural land. Rural communities slowly found themselves part of the growing Bangkok metropolis, but without the infrastructure necessary to survive in a city. For instance, most villages were (and still are) unpaved, unsuitable for cars or larger vehicles, and reliant on the klongs for fresh water. Without agricultural work, former farm labourers either had to develop new business skills, or take unskilled work in factories. Thai development has focused largely on economic rather than social development, and has not extended to urban redevelopment as a rule.

Modernisation of the political system lagged behind economic and social change. The country remained an absolute monarchy until 1932 when a revolution led by young army officers, unemployed foreign-educated youths and some older moderates replaced the monarchy with a more western-style constitutional parliamentary system. The new system was dominated by the military. It also retained the traditional Thai patron/client power relationship – a feature that is still strong in modern Thai government.

In 1938, Field Marshal Plaek (known as Phibun) took power and became

Prime Minister. His term of office was marked by a policy of nationalisation, although the policies were seen as a continuation of the modernisation of the monarchical system. During this time the western calendar was adopted, and Thais were required to salute the flag, know the national anthem and use the national language rather than regional dialects. People were encouraged to buy Thai-only products (a policy that discriminated directly against the Chinese population, who at that time were a major economic force in Siam). People were encouraged to dress in the western fashion – men in coats, trousers, shirt, tie; women in skirts, blouses, hat, gloves; and all in shoes – and to wear hats when entering government offices. A husband was supposed to kiss his wife before going to work. These policies were seen as necessary for progress and civilisation so that the world would see the country as a modern nation (Wyatt 1984). The country's name was changed from Siam to Thailand in 1939.

After the Second World War, the Thai economy changed again as the European influence expanded in the new countries of post-colonial South-East Asia. Farmers changed modes of production to grow cash crops, and the 'green revolution' was introduced to stimulate a new agricultural plantation system. The commercial sector was marked by increasing specialisation and compartmentalisation.

At the same time, there was widescale adoption of western models, in government, industry and society generally. For instance, roads increasingly replaced klongs as Bangkok's transport system. The major impetus behind the move from klongs to concrete highways was the use of Bangkok as a military base by the Americans during the Vietnam War. A western-style government bureaucracy grew up to administer the country, although it retained aspects of the traditional Thai system of hierarchical authority and organisation.

The transition from a closed and self-sufficient economy to a more open system firmly tied to western models of production and tied to the international economy quickened significantly, particularly in the 1960s (Figure 8.2).

Bangkok Today

Bangkok owes its dominance to three factors:

- geography;
- the concentration of economic activity; and
- the centralisation of political and administrative functions,
(Phipatseritham 1983)

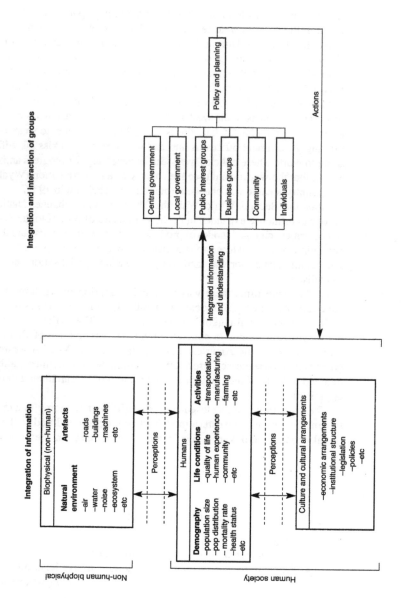

Figure 8.2 Integration in the study of an urban ecosystem

Although the prime reason for establishing Bangkok as the capital was defence, its location near the mouth of the Chao Phraya River was ideal for carrying out trade. The Chao Phraya drains a large portion of the country and has served as a natural highway from the hinterland to the sea for centuries. Thus much of the country's agricultural wealth flowed through the burgeoning port – to the exclusion of almost all other Thai seaports.

As a result of the flow of products through the city, Bangkok developed as a centre of commerce for agricultural products. Later the increasing wealth in the city encouraged the development of industry and manufacturing. Railways, roads and a port (and later highways and an international airport) were built to facilitate the country's economic activities. Since the 1960s, the bulk of Thailand's manufacturing and industrial activity has developed in the city. By 1990, Bangkok had over 20,000 factories.

After the Bowring Treaty of 1855 transformed the closed Thai system to a more open economy, most of the international trade was channelled through the port of Bangkok.

After the 1932 revolution, the military consolidated the country's administration. Government in the provinces was transferred from local hereditary rulers to Governors appointed by the central government in Bangkok. The centralisation of political, administrative, military, economic, educational and social power in Bangkok has increased since the 1960s. The massive centralisation of Thai government has prevented the development of genuine representative local government until very recently (Dhiravegin 1983).

As well as the 7–8 million residents of the city of Bangkok, up to two million more commute to the city every day for work. Hundreds of thousands of Bangkok residents live on or below the official poverty line, living in slum areas dotted throughout the city. The cost of living in Bangkok is significantly higher than in other parts of the country.

The city layout is a fascinating mix of residential and commercial building styles, all in transition. Along major roads are 'shop-houses', with shops or small factories for hand made goods on the ground floor, and rooms for the members of an extended family on between one to four floors above. Those which have been converted to small factories make everything from printed T-shirts to doing motorbike repairs. The activities of the shops and factories often spill onto the pavements.

Behind the orderly rows of shop-houses are communities of mixed housing, entered through *sois* or lanes. Since these are laid out by residents and the original developers, and are not subject to planning or regulation, their routes and widths are quite irregular. The housing and activities are mixed – the homes of the rich and relatively poor may be juxtaposed, and small service industries are interspersed among the houses. Some communities are relatively homoge-

141

neous, depending on the ownership of the land they occupy and its cost. It is common for people to build their own houses, on land owned by temples, the government, or private owners. They are now vulnerable to eviction as rising land prices prove attractive to the owners.

Slums are prevalent. Some are permanent and being assisted to negotiate for land tenure. These are being provided with financial support to upgrade their infrastructure, and eventually their standards of housing come to appear little different from other low to middle income groups. These slums receive the assistance of non-government organisations (NGOs), and may be quite effective politically. Others are very new or temporary, built on vacant land available briefly.

In the newer areas of the city, or inner areas which have been cleared for rebuilding, are housing estates designed for various income groups. These are particularly prevalent in the disappearing western green belt.

There is no single 'central business district'. At least four areas of high rise office blocks and hotels punctuate the Bangkok skyline, some mixed with high rise apartments. Former business districts, close to the older quarters of the city, are characterised only by their density of shop-houses and the occasional modestly tall department store.

Bangkok is massively polluted. Air pollution is chronic, although official statistics disguise its severity. Inadequate sanitation and policing of regulations means that raw sewerage and industrial waste is pumped straight into Bangkok's klongs. Dissolved oxygen levels in the Chao Phraya River are virtually negligible, which, combined with high levels of industrial and organic pollution, make aquatic life impossible.

Bangkok Planning

The government's modernisation policies since the 1960s have focused on the provision of infrastructure, such as roads, but have left economic development within the hands of private enterprise.

One result of these policies is that the construction of new roads leads to land speculation and changes in land use. One example of these processes in action is the growth of urban areas in the 'Green Belt' that surrounds the city. New roads, constructed by the government, allow middle-class workers to escape the congestion and pollution of the inner city, but still commute to work every day. Demand for land and housing in the new housing estates leads to rising land prices and a housing boom. The original residents, often agricultural workers or poorer people, are usually unable to afford to live in the area and are forced out. Some end up living in slums in poorly serviced areas of Bangkok. Because of the distance of the new housing estates from the city centre and workplaces, and the lack of public transport services, the new arrivals are forced to buy cars, further adding to the city's traffic load.

The 'Green Belt' itself was an attempt to protect the agricultural land surrounding the city from urban sprawl and the expansion of industrial and commercial activities. The government gave these areas legislative protection in 1975 and 1979. The legislation permits only a limited proportion of construction within the Green Belt zone, and that must be related to agricultural activities. However many government agencies, industrial investors and land developers believe that the Green Belt policy is a major obstacle in the way of the city's economic development. There have been many arguments over land use within the Green Belt (Karnchanapant, 1989). Meanwhile so many land developers have been able to capitalise on disagreement and circumvent the legislation that the green spaces are disappearing rapidly.

These two examples show the piecemeal approach to planning in Bangkok. There have been several attempts at comprehensive planning for Bangkok in the last 35 years, but none has been successful. The majority of the plans have emphasised the development of western-style infrastructure for the city, such as the development of roads to replace existing 'traditional' modes of transport such as the klongs. None of the plans have been implemented, although some aspects have been implemented in isolation. To a certain extent, all of the comprehensive plans can be seen as part of an attempt to modernise and westernise the city.

The majority of urban development occurs as the cumulative effect of businesses and residents' actions, conducted in a *laissez-faire* policy context. Efforts to correct urban problems have focused largely on providing physical and engineering infrastructure, rather than social or integrated solutions to Bangkok's environmental problems. Urban reform has been neglected apart from occasional limited initiatives, such as a burst of construction of flyovers to alleviate traffic congestion at major intersections in the late 1980s, and long-term community development work by the National Housing Authority and non-government organisations (with international support and local leadership) in slums.

An important feature of Thai government planning is that most regulatory instruments have built-in loopholes – for instance, in some areas, 10 per cent of the buildings may breach the building regulations. Since no-one specifies which 10 per cent, breaches quickly exceed that limit and attempts to regulate building activity are deemed a lost cause. Similar features can be found in all other aspects of urban planning.

Because of the lack of effective control over land use, urban expansion is taking place without sufficient planning, thus aggravating the inadequacies of the transport network, and causing further traffic congestion in all parts of the city.

Bangkok's Transport System

One of the changes brought about by the westernisation of Thailand has been the change in the transportation system.

Traditionally, Thais have used water for most of their transportation. Pre-1855 Bangkok, as well as much of lowland Siam, was criss-crossed with klongs and natural waterways, which were adequate for most transportation needs. At the turn of the century, Bangkok's klongs earned the city the title of 'Venice of the East'.

In the drive to modernise Thailand, and particularly Bangkok, the traditional water-based transport system was replaced with road transport on the western model. Since the 1960s, many of Bangkok's klongs have been filled in and new roads built. Those that remain have been allowed to degenerate into little more than unregulated sewers.

Apart from changing the city's traffic patterns, the loss of the klongs has seriously affected the city's drainage and flood control system. Some areas remain flooded for several months every year, and the loss of the natural flushing of the city's waterways has contributed to increasingly unhygienic conditions, particularly in the lower income areas.

Road construction is largely the responsibility of central government. However, because of the nature of Thai government and the number of agencies involved, the system for planning and constructing roads is very inefficient. Road surfaces are not expanding quickly enough to cope with the great increase in the number of vehicles. In 1990, the total length of roads – major, minor and access roads – was 2800 km. This provided Bangkok with a traffic surface area of 38.44 sq km in a city of 1568 sq km (2.5 per cent of the total Bangkok area – the recognised standard for western cities is about 20–25 per cent).

In 1993, the city had two and a half million registered vehicles, compared to 1.3 million in 1986. There are now at least 600,000 cars and 890,000 motorcycles. The city could have about four million cars by the year 2000 – one car for every three people living in Bangkok, or 1.8 for every household. These figures do not include those vehicles from surrounding provinces which commute to the city each day.

Traffic congestion is continuous throughout the day, with little change at peak hours. The average speed on main roads is about 13–15 km/h, reaching 20 km/h on some larger roads. A recent study measured the average speed at 7.7 km/h inside the middle ring road, and as low as 3–5 km/h in the seriously congested central business area. The study predicted that the average travel speed would be 5 km/h by the year 2006 if no remedial action was taken. It is estimated that the average Bangkok resident spends a total of more than 40 days a year in Bangkok's traffic (JICA, 1989).

The range of engineering solutions available to relieve Bangkok's transportation problems is limited. Underground road and rail systems, such as those common in Japan and many western cities, are feasible but extremely expensive.

Rail has never been strongly developed in Thailand. The few rail routes interfere with the flow of road traffic where roads and railway lines cross, especially at peak hour. Overhead rail systems, such as a 'skytrain' are under development, but because of the expense involved, only a few routes are possible. Three government departments have planned road and rail rapid-transit systems, but with no coordination between the three projects.

There has been a revival in the use of the klongs for transport. However because many klongs have been filled in over the last thirty years, their loss limits the effectiveness of water transport in modern Bangkok. Also, the diesel engines which power the klong boats contribute to noise and air pollution. Severe water pollution makes this form of travel an active health hazard. Many of the boats operating on the klongs are entirely covered to ensure that commuters are not splashed with polluted water. Wave action also damages the klong banks and foundations of buildings built on them.

At present, the main mass transportation system is the state-run public buses, with rail used increasingly on the routes available. Although Bangkok has about 7000 buses, they are not coping with the growing population. The fleet is old and the exhaust fumes from the buses contribute seriously to air pollution. At any one time, only about 70 per cent of buses are in use. The Bangkok Mass Transit Authority, which operates the buses, is under pressure from heavy debts since the bus fare is controlled by the government, which keeps the price very low. The management of the system is also very inefficient. Consequently, this sole form of public transport provides poor service and does not help relieve the transport situation (Figure 8.3).

Stakeholders in Bangkok's Transport System

Parliament and Politics

The Thai parliament was established in 1932 following the revolution which replaced the monarchy with a constitutional parliamentary democracy. The constitution itself has been amended 17 times since 1932, as Thailand has lurched from one political crisis to the next. Military coups are frequent and there is a strong military involvement in civilian government.

Parliament consists of two 'Houses' – a lower house consisting of directly elected members, and an upper house consisting of appointed senators – mostly from the military, senior bureaucracy and the business sector. The Thai Upper House is largely an advisory and cautionary body. It has the power to slow the passage of legislation proposed by the lower house, but it cannot block it.

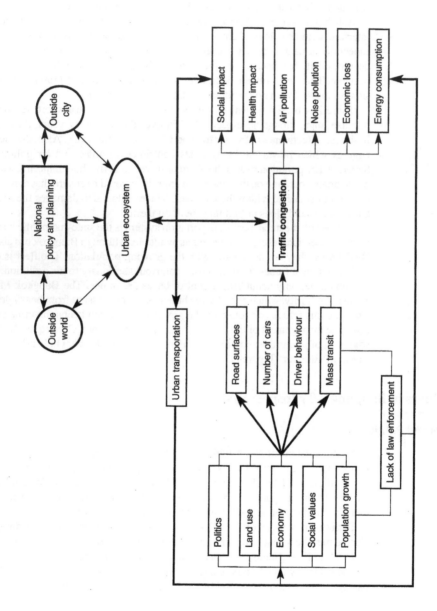

Figure 8.3 Traffic congestion: some important interrelationships

One important feature of Thai politics is the number of political parties involved. Virtually every elected government (as opposed to military appointed ones) in the last two decades has been a coalition, usually with a minimum of three parties, each with a different social agenda.

Thailand's multi-party political system, and the vulnerability of coalitions amongst the parties, means that it is hard to reach agreement on a policy or proposal and stay in power long enough to implement it. Elected governments have become paralysed with indecision about who to please, especially in the absence of a clear electoral mandate for a particular set of policies.

Another feature of Thai politics is the almost continuous series of coup attempts. As a rule they are usually bloodless affairs, with only three major exceptions in the past twenty years: the crushing of demonstrations in 1973, 1976 and 1992. Politically, Thais have been moving increasingly towards directly accountable government, and are showing increasing intolerance for military intervention in civilian government.

In 1992, 50 people died as the military ended weeks of prodemocracy demonstrations, protesting the appointment of a non-elected General as Prime Minister. The general resigned after the crackdown, making way for fresh elections. Those involved in the protests came from every section of the Thai community, showing that the general public had developed minimum expectations of democratic government, and was prepared to fight for them.

On the other hand, Thailand has made significant progress while under recent military rule. The most reformist of recent Thai governments was appointed by the leaders of the 1991 *coup d'etat*. The Prime Minister and Cabinet – both appointed by the military – pushed through an astonishing number of sensible and obvious reforms in areas such as health and safety (such as making the use of motorcycle helmets compulsory). After this experience, many Bangkok residents favoured full democracy in principle, but appreciated the efficiency of this appointed leadership in practice.

Thai democracy has now matured to the extent that politicians are beginning to respond to their electorates, rather than to vote-buying and the social and economic elite of business, military and hereditary interests.

The elite remains a potent force in Thai society however. Politicians, business figures and senior figures in the military and bureaucracy are linked by family and marriage, and business interests. One result of this arrangement is that threats to one sector affect other members of the elite.

Another problem facing elected coalition governments is the appointment of ministers. Each party vies for ministerial posts, because they are the key to implementing political agendas. The result is that ministers compete with one another not only on policy issues, but on political lines as well.

Government Agencies

Government administration in Thailand is highly centralised, and until recently, there was little role for local government. As a result, central government departments are responsible for not only national policy (eg defence, trade and economic planning), but also for issues usually allocated to local governments in the West (activities such as local road building and provision of infrastructure).

The structure of government departments changes infrequently, and the organisation of responsibilities between levels of government and between departments remains somewhat arbitrary. Powerful departments, and central government as a whole, are reluctant to give up their responsibilities in the interests of rationalisation. For example, the Department of Industrial Works (DIW), the Office of Town and Country Planning (OTCP), the National Economic and Social Development Board (NESDB) all compete with each other and the much less influential Bangkok Metropolitan Administration (BMA) for planning control over the city. Some urban functions (such as transport) involve a myriad of other government agencies. For the BMA to carry out all of its duties properly, it has to liaise not only with the above but also with the Traffic Police Division, Ministry of Education, Ministry of Public Health, Ministry of Transport and Communications, Ministry of the Interior, the Metropolitan Water Works Authority, National Housing Authority, Bangkok Mass Transport Authority, Express and Rapid Transport Authority, Telephone Organisation of Thailand, and the Industrial Estate of Thailand (United Nations, 1987)

Up until very recently, communication among departments used to follow hierarchical lines, with individuals in different departments unable to communicate directly with their opposite numbers. All communications had to travel to the top of the initiating organisation's hierarchy, to be sent between heads of departments, then filter down to the action level in the other department. Responses travelled by the same route. More recently however, the public service has recognised the need for coordination of related activities in different departments. Because the bureaucratic structure has proved difficult to change, a system of coordinating committees has been grafted onto the system. Although this strategy offers the possibility of increasing levels of interaction and the exchange of ideas, so many organisations need to be involved in each committee because of the way that responsibilities are divided, that active coordination is a nightmare. There also need to be innumerable committees, with the result that public service managers can find themselves tied up in an extraordinary number of them. Some 30 organisations and committees are involved in Bangkok's transport. Individuals and departments determined to make reforms in their areas of responsibility can make only slow headway under these arrangements.

Good administration is also hampered by competition between government departments. A classic example is the set of proposals for rapid transport: three organisations have three separate projects, each featuring different modes of transport. The State Railways Authority's 'Hopewell' project features an elevated road and railway, above existing transport routes. The Bangkok Transportation System's 'Skytrain' involves elevated light rail, and the Ministry of the Interior has a major expressway project. Because each proposal has been developed in closely guarded isolation, it will be extremely difficult to integrate all three into an effective transport system. Since this development, Bangkok has commissioned a mass transport master plan, which recommends that part of each project be routed underground.

Most government policy has focused on development through economic growth. Most economic and social planning (including aspects of education and health) have been directed towards this end. Thailand's economic and social planning is in the hands of the National Economic and Social Development Board. The Board's planning is largely at the policy level however, and provincial governments are expected to interpret how they should respond to the policy blueprints.

The NESDB is enthusiastic about the prospect of Bangkok becoming a mega-city (a prospect that leaves local authorities such as the BMA aghast). The result for Bangkok is that any urban infrastructure planned now, especially sewerage and transport infrastructure, will be competing financially with infrastructure necessary for more direct economic purposes.

Central planning has had its benefits for Bangkok however. It has ensured services essential to the country's trading performance – such as port facilities and transport – have been constructed.

Bangkok Residents

Some idea of the state of life for the average Bangkok resident can be gained from health, population and urban statistics. Overall, they paint a picture of increasing stress and disruption for many Bangkok Thais.

Population growth in Bangkok is the consequence of two forces – natural increase and migration from other parts of the country. When the city was established in 1782, it had a population of about 50,000. The population size increased only slowly in the eighteenth and nineteenth centuries, but increased dramatically in the twentieth. By 1988, there were over six million people registered as living in Bangkok. The actual population is believed to be much higher. In 1989, the National Statistical Office reported that 30 per cent of people living in Bangkok are not registered as citizens, putting the city's population nearer to eight million – excluding the two million daily commuters from neighbouring provinces.

The population of the Bangkok Metropolitan Region (BMR) has been growing at an average of 3.8 per cent per year for the last 25 years. According to Thai census results, the ratio of the city's population to the total national population is also steadily increasing, indicating a net movement of people into the city. Indeed, Bangkok's high population growth is due mostly to immigration from other provinces: the rate of population increase due to births has been falling in Bangkok for several years. It is estimated that by the year 2000, Bangkok will have a population of 8 million registered citizens; the Bangkok Metropolitan Region, 11 million; while Thailand's total population will have exceeded 64 million.

Bangkok's population amounts to over ten per cent of Thailand's total population. They live in an area of about 1600 square kilometres – only 0.3 per cent the size of Thailand. As a result, population density is very high. In 1988, there were 3646 people living in each square kilometre in Bangkok, compared with an average of 91 people per square kilometre for the rest of the country.

Thais are attracted to Bangkok from other parts of the country because of its high industrialisation (and therefore good job prospects) and extraordinary centralisation (and access to services and amenities unavailable in the provinces). The largest source of immigrants to Bangkok from the provinces is the poorest northeast region.

The impressive growth that has seen Thailand dubbed one of the five 'Asian Tigers', has been associated with a widening gap between the rich and the poor, as well as between the urban and rural sectors. By 1985–86, the richest 20 per cent of the population received over 55 per cent of the country's total income, while the poorest 20 per cent earned less than 5 per cent. There is a twelve-fold difference in income between the richest and poorest 20 per cent of the population and the gap between rich and poor continues to increase (Bhongmakapat 1990; Hutaserani and Jitsuchon 1988).

Table 8.1 below shows poverty levels in Thailand. In 1988, more than 25 per cent of all Thais lived below the poverty line. If the figures are further analysed, the statistics reveal that 30 per cent of rural Thais live below the poverty line, compared with five per cent of urban Thais. Although the proportion of the population living below the poverty line has decreased since 1988, the percentage of people living below the poverty line remains high among those that live in the suburban and urban fringes of the city, particularly farmers, gardeners and slum dwellers.

Table 8.1 Percentage of the population under the poverty line

	1975–6 (%)	1980–1 (%)	1985–6 (%)	1988–9 (%)
Thailand	30.0	23.0	30.0	24.0
All villages	36.2	27.3	35.8	29.4
All sanitary areas	14.8	13.5	18.6	13.2
All municipal areas	12.5	7.5	5.9	7.0
Bangkok Metropolitan Region	7.8	3.9	3.5	-
BMR – city core	6.9	3.7	3.1	3.3
BMR – suburbs	6.0	2.6	2.5	1.6
BMR – fringes	12.0	9.2	8.8	6.3

Source: TDRI (1993)

Health statistics provide an interesting profile of Thai living. For Thais generally, the life expectancy has increased since the 1950s. At present, men can expect to live for 62 years and women 68 years. The figures are expected to reach 67 years for men and 71 years for women by the year 2000. In the last 40 years, the death rate has halved. Infant mortality has also fallen (Division of Health Statistics of 1989). However, the differences between Bangkok and other parts of Thailand are pronounced. In 1987, there was one doctor for every 1418 Bangkok residents; for the rest of the country there was one doctor per 8871 Thais. Other health professionals such as dentists, pharmacists, nurses and midwives are also heavily concentrated in Bangkok. There is one hospital bed for each 316 Bangkok residents – in the rest of the country the figure is one bed per 855 persons.

But the fact that the population of Bangkok has better access to medical services than people in other regions does not mean that they are living under healthier conditions. The proportion of deaths in Bangkok caused by heart disease, accidents and cancers is much higher than the national average. In the case of heart disease, the figure is almost twice as high. All three diseases are related to urban development and industrialisation (Division of Health Statistics of 1989).

A 1986 study of 2000 households revealed that 44 per cent of Bangkok residents showed some symptoms of psychological problems – including stress, sleep disorders and mental fatigue. The figure for those over 16 years of age was 49 per cent. People suffering from psychological disorders came mainly from the less-educated, divorced and low-income groups (Meaksupa et al 1987).

Bangkok's crime statistics are alarming. In 1988, there were 84,127 crimes reported for the whole country: of these 20,689 (24.6 per cent) were reported in Bangkok. There were more than 2000 violent crimes – including murder, robbery, arson – committed in Bangkok that year. Police estimated that the rate of violent crime in Bangkok was 37.7 incidents per 100,000 people – or one violent crime reported every four hours. The murder rate was also very high: 6.1 per 100,000 people, or one murder every 25 hours. Levels of rape reported to police were about the same as for murder. Robberies were the most common violent crime, with one committed, on average, every 38 minutes. The actual incidence of violent crime might well be higher than the figures reported by the police (Police Department 1990).

Drug addiction also remains a deep concern to both the authorities and the community. The BMA's Department of Health reported that about 46,000 drug addicts entered Bangkok's hospitals in 1988. There were about 3000 new drug addicts reported every year.

Bangkok has very little land for recreation – it has only eight major parks between an estimated six million people. The parks amount to less that 0.2 per cent of the metropolitan area: this equates to only 0.43 square metres per person living in Bangkok. Even factoring in all other green spaces in the Bangkok Metropolitan Area such as the 'Green Belt', the figure rises to only 1.90 square metres per person. This is a low figure compared to other major cities:

	sq m
• London	30.4
• New York	19.2
• Montreal	13.0
• Paris	8.4

More than half of the people that make use of Bangkok's parks are the urban poor (Taweesuk 1990).

The high concentration of all kinds of economic activities in Bangkok has led to intense land use around the inner city area. The increasing demand for space for commerce, business and accommodation forces up the price unrealistically and increases the concentration of buildings. This has forced out many middle and low income earners. They have had to move toward the outer edges of the city. Although they are compensated in part by the lower land prices, more plentiful housing and less environmental degradation, they are condemned to spend significantly more time in Bangkok's traffic.

Not everyone has given up living in the city centre. The city has seen a massive increase in the amount of high-rise living in recent years, particularly in the inner city. In 1989, there were 968 buildings with more than six storeys, of which 29 per cent were used for accommodation – mostly flats, apartments, condominiums, courts, dormitories or guesthouses. There were 1172 high-rise building project proposals made in 1988 alone, with another 1397 proposals expected for submission in 1989. The increase in inner city living has led to increasing demands on garbage collection, wastewater treatment, and public transport.

Another growth area in Bangkok's housing environment are the slums. A study in 1987 found 1500 slums in Bangkok. Altogether there were 235,000 slum households reported by the study, housing a population of about 1.3 million people. Many people living in slums have been displaced from inner city areas or agricultural land, or were new arrivals from poorer outlying provinces. Slums are not confined to the city centre, but can be found in all parts of the Bangkok Metropolitan Area.

The number of slums has increased dramatically in recent years despite official attempts to reduce their number. Closing down slums or evicting squatters carries with it the threat of violence and the intensification of social conflict.

Buddhists and Buddhism

Thai Buddhism has a mixed role in Thai society. Some parts serve to maintain the *status quo*, but others inspire public support for a variety of issues, causes and values.

Buddhism has been in decline in Thailand, particularly among the young and those in higher income groups. But despite the decline in religious observance in Bangkok, the *Wat* (Buddhist temple) remains the focus of many neighbourhood communities. Different income groups share religious observance at the same Wat, and 'merit-making' (donations to Buddhist temples) is a prestigious activity for all income groups. Wats are more common in the older sections of Bangkok, and are few and far between in the new housing estates. The Wats have traditionally had an educational role within their local communities. Buddhist monks travel widely within the community and are highly respected. They promote a strong vision of a society based on Buddhist values which involve, in part, respect for the environment and peoples' rights. However, Buddhist values, which are based on traditional agricultural Thai life, are increasingly foreign to many city-bred Bangkok residents. Some see the Buddhist influence as antithetical to the industrialisation of Bangkok and Thailand.

The Buddhist tradition of respecting all forms of life has resulted in monks taking an active and prominent place in Thailand's environmental movement and in rural community development initiatives based on sustainable farming practices and conservation. A few Buddhist intellectuals and social reformers achieve influence through their capacity to communicate at all levels of Thai

153

social hierarchy. Some have excellent communication with the Royal Family, yet listen to and represent the concerns of those with low social and economic status. These individuals are prominent in a network of non-government organisations (NGOs) which link environmental and human rights concerns. They have also proved to be impervious to types of pressure which silence other critics of government – particularly in periods of military rule. They have been visible both as organisers and among groups that have protested military intervention in Thai government. In addition, Buddhists have helped a great deal to inspire and strengthen the non-government organisation (NGO) movement in Thailand. The NGO movement in turn has been focal in new oppositional alliances against unpopular development proposals such as a series of dams.

Thai Buddhism and its code of generosity and tolerance are under enormous pressure in Thai society. One cause is Bangkok's traffic problems, which lead to selfishness and frustration on the part of drivers and bus passengers. On the other hand, meditation techniques, and cultural values towards remaining calm, help people to cope with their stresses.

The Traffic Police

Senior officers have long been frustrated by the side effects of engineering efforts to improve traffic flows, the typical response of their political masters and other government departments. The police have been critical of isolated roadworks projects such as flyovers and new freeways because most are poorly integrated with other roads near the entry and exit points, they cause enormous disruption during construction, and in the absence of efficient public transport they encourage more private cars onto the roads.

Traffic police, who are more exposed to traffic and air pollution than travellers, are now protected with masks, and have access to respite booths and oxygen tanks on the worst intersections. Nonetheless, it is estimated that 20 per cent of Bangkok's traffic police suffer from heart conditions directly related to air pollution.

Traffic police also suffer from enormous stress. Newspapers carried the story recently of one officer who became so disillusioned and frustrated directing traffic on one of the city's major intersections that he changed all of the traffic signals to green and danced among the ensuing chaos.

Industry and Business

Industry and business are frequently painted as the villains in Thailand's growth – the major beneficiaries of Thailand's drive to modernise and industrialise. Businesses are very diverse: from multinational companies to small family-owned businesses run by people from low-to-middle income groups. Forty per cent of Thailand's registered factories are in the Bangkok Metropolitan Area, and 60 per cent of these are small businesses. There are also many illegal factories. Exploitation of Thailand's cheap labour force, polluting the environment with impunity, bribing local politicians and not being concerned with the health and well-being of Bangkok residents are all crimes that businesses – particularly international companies – have been accused of. There appears to be a strong business influence on the creation and exploitation of planning loopholes. The building environment of Bangkok shows many examples of powerful business organisations overcoming public service attempts to regulate building and land-uses, through personal contacts with Ministers, or financial influence at decision-making levels. On the other hand, some businesses have been champions of reform and restructuring.

One example of a reformist business group is the Thai print media, which is privately owned. Despite significant regulation of the material they may publish, newspaper owners have been courageous critics of all governments. Newspapers are a potent force in encouraging political accountability and public participation in the democratic system. The media has also been vital in publicly-initiated protest campaigns and in shaping new values in Thai society.

While the business community has responded strongly to the government's drive for economic growth, many individuals remain at least partly motivated by Buddhist principles. 'Merit-making' at temples is still a strong social demand, and some business people allocate part of their wealth to social causes. Business people cannot be characterised as predominantly motivated by financial gain.

Businesses suffer from the poor management and planning of the modern Bangkok. They lose millions of working hours every year due to traffic congestion. Holding meetings is difficult because it is impossible to predict when people from other parts of the city might arrive. It makes transporting goods and materials across the city or to the port a nightmarish operation.

Businesses also pay for Bangkok's poor road system through lower productivity, and a greater exposure to accidents and mishaps, caused by employees that arrive at work tired, stressed, aggressive or simply worn out after several hours in the traffic.

Problems Caused by Traffic

Economic Losses

The amount of time wasted in traffic congestion and the poor public transport system encourages the growing middle-income population sector (who can afford them) to acquire private cars. In addition, those who move to the new fringe areas of the city, often in advance of public transport being supplied, depend on their own cars. Because of the lack of effective control over land use, urban expansion is taking place without sufficient planning, and the inadequacies of the public transport system are aggravated.

The opportunity cost of time wasted by commuters on the road has been estimated at about 23 million *baht* per day, and the cost of fuel wasted is up to up to 5 million *baht* per day. Altogether, the economic loss from the mismanaged transport system is more than 10,000 million *baht* per year.

Pollution

Traffic congestion is one of the main causes of air pollution in Bangkok. Airborne solids, noise pollution, lead levels and carbon monoxide levels are all at serious levels, and increasing.

The official monitoring system obscures air quality status, since the Thai standards are based on average levels of pollution over specified periods (1–24 hours, depending on the pollutant). The WHO recommended standards are framed in terms of thresholds which should not be exceeded once a month or year. The Thai average may sit comfortably under the threshold line, but actual readings may cross the threshold frequently. Some of the Thai standards are also twice as generous as the WHO ones. The monitoring stations are also located in areas which are safe from interference, which means they are also well away from main roads where readings are likely to be highest.

A US study carried out in 1990 ranked lead and suspended particulate matter (SPM) as two of the three highest environmental risk problems in Bangkok (the third was disease caused by infections or parasitic organisms). The same study ranked airborne carbon monoxide (CO) and metals other than lead as a medium level risk, while the low risk category contained problems posed by toxic air pollutants such as sulphur dioxide, nitrogen dioxide and ozone. In Bangkok, a major portion of the pollutants comes from burning petrol and diesel in cars and buses (Abt Associates and Sobotka and Co 1990).

Suspended particulate matter (dust) levels ranged from just under the Thai limit to double that limit in 1983–1986, then increased to up to twelve and a half times the limit by 1989. Suspended particulate matter levels are highest in the morning when traffic loads are heavy and the air is calm. Particles affect visibility and may be toxic. SPM irritates lung tissues causing long-term breathing disorders or cancers.

Lead levels are increasing despite a decrease in petrol lead levels because of the rate of increase in vehicle numbers. Maximum lead levels increased five-and-a-half times between 1986 and 1989. High concentrations of lead in the body can affect brain, liver, kidney and reproductive functions; limit the formation of blood cells and generally interfere with the growth of tissues in the body. Children are particularly sensitive to lead poisoning.

The main source of lead in Bangkok is from leaded petrol and diesel. It may be absorbed directly through exposure to polluted air, or indirectly through food that has been contaminated during transport on Bangkok's roads, or as it has lain in street-side stalls and markets.

Carbon monoxide readings for a one-hour average range as high as 60 milligrams per cubic centimetre, compared with a WHO threshold of 25 milligrams to be exceeded no more than once a year. A 1988 study found that the most significant risk of carbon monoxide (CO) poisoning was among those that lived or worked in buildings close to busy streets. Carbon monoxide poisoning has been linked to loss of productivity among workers and to general discomfort. It can affect the central nervous system, and may affect physical coordination, vision and judgement. It can also affect the cardiovascular system, and make existing heart conditions more serious. Carbon monoxide is known to affect hundreds of thousands of Bangkok residents.

The three factors that lead to air pollution caused by traffic are:

1   the traffic levels;
2   the lack of regulations requiring all motor vehicles to have annual inspections; and
3   the weakness of law enforcement.

Measures of noise levels at road sites in Bangkok are frequently above 80 dBA (a weighted scale of decibels). The internationally accepted level is 70 dBA (Kangagate, forthcoming)

A study found that 21 per cent of motor vehicles had noise levels above the Thai standard (100 dB at half a metre from the exhaust – the equivalent of a pneumatic drill). Eighteen per cent of trucks and 15 per cent of tuk-tuks (three-wheeled motorcycle cabs) also exceeded the noise standards (Department of Land Transport, 1990). Eighty per cent of klong boats surveyed in 1983 had noise levels above 80 dB, with 80 per cent of operators having hearing loss.

The major causes of vehicle noise include:

● tampering with exhaust pipes to deliberately increase the noise level (popular among motorcycle owners);
● failure to replace worn exhaust systems;
● inadequate enforcement of regulations;
● inconsistencies between different government departments in adopting noise standards;

- inadequate funding for checking noise levels; and
- a lack of public awareness of the problem and willingness to cooperate.

Noise pollution affects sleep and relaxation, resulting in fatigue and mental stress, and affects the digestive and cardiovascular system. Noise also makes talking and communicating more difficult, affecting social relationships.

Impacts on Society

In a recent study, residents were asked what they saw as the Bangkok's greatest environmental problem (Poungsomlee 1991). Traffic congestion was the first response of almost all of those asked.

> All around the place there are traffic jams ... and during the rush hours there are always traffic jams with lines of vehicles unable to move. It might be quicker if one walked to one's destination ... there are jams most of the time.

Bangkok residents were well aware that the roads are unable to carry the number of cars on them. Even the expressways are insufficient to cope with the problem.

> Expressways are likely ordinary roads ... they are no longer 'express' ... we are charged [for using them] too. Although many expressways are to be built, they will not solve the problem.

Residents were also well aware of the connection between traffic and air pollution and noise:

> Even the trees are dead ... [they] cannot bear it. I doubt the idea of growing trees to minimise air pollution is effective. The trees now standing are dying.

Those most at risk from air pollution are people in low income groups who depend on motorcycles and buses for transport. They are exposed for long periods to air pollution caused by traffic. People in higher income groups have air-conditioned cars (if they can afford them) to protect them while they are travelling, and air conditioning at home.

People unable to afford a car, but unwilling to use the slow and overcrowded public buses opt for motorcycles. Motorcycles fill an important niche in Bangkok's transport system because of their ability to enter narrow *sois* (lanes) and to weave around heavy traffic. They are cheaper to purchase than a car, cheaper to run, and cheaper to hire. Despite being illegal and dangerous, hiring is popular. The high-pitched noise of motorcycles is the most irritating and disruptive to city residents.

Complaints of ill health due to air pollution caused by traffic congestion are common. The most commonly reported problems are headaches, eye irritation, infected lungs and heart, and feeling weak.

Traffic also causes great stress, which in turn may damage personal relationships. People become irritated easily and lose their tempers.

> ... the most obvious evidence [of psychological problems caused by the traffic] is when one is going to drive a car. Then one changes his personality to be aggressive ... one of the school teachers bites a handkerchief every time he drives a car to prevent himself cursing others.

People living in different parts of the city cope with the city's poor transport system in different ways, depending on how far they have to travel, when and where they have to work, and their mode of transport. Even if they have only a short distance to travel, many people may leave home as early as five am to ensure shorter and more reliable travel times, or to secure parking space. In order to travel early, some people start work up to two or three hours before their official starting time and remain longer at the end of the working day to avoid the heavy traffic on the way home.

> Later than 6 am, there is no parking around the Ministries area [in the central business district] .... In government offices, a traffic jam cannot be used as an excuse for being late .... If there is a traffic jam, you should get out earlier .... Some factories cut off part of the wage if workers are late. My daughter is one. So she runs to work if she is going to be late.

Another consequence of long travelling times is that people have little time to relax at home, spend time with their families, or to carry out their household responsibilities. Meals and preparation for the day are rushed, or are carried out on the way to work. It is not uncommon for Bangkok residents to feed their children breakfast and supervise homework in their cars. On weekdays, the long journeys to and from work are the only chance that many working people to have time to talk with each other.

> I have my children do their homework at the university [the work place] or have them tutored there. We have to have everything done out of the house. Then when we reach home, we just take a bath and go to bed .... Everything is done in cars. If we could move our beds into cars, we might do that!

Because they are so stressed and tired when they arrive home, most people do not feel like interacting with other members of the family. Many Thais living in Bangkok are concerned that the quality of relationships within the family is declining, partly as a result of the traffic.

> There is a traffic jam, so we come home irritated and quarrel with family members .... Sometimes my wife is the scapegoat. When I come back from work and it gets very hot, I put the blame on her...

Many parents are concerned about the limited amount of time that they have to give to their children because they have to go out early and do not get home until late.

> ... when the father leaves home, the children are still asleep. When he comes back home in the evening, they have gone to bed. That's not a very good relationship. Children are closer to the mother. Well they understand that their father goes out to earn some money so that they can learn [to pay education expenses], but they are certainly less close .... One of my friends at work said that he has to get up at 4 am to cook rice when his child is still asleep. He said that sometimes he has to sneak out of the job to see his child. His daughter asks where he has been. He doesn't go anywhere, just goes to work.

As a result parents are concerned about how their children are growing up, and that they can not control their children. Another worry for many already overworked parents is that they can not prepare suitable food for their children because they have to hurry to work. When the children wake up, they eat whatever they want.

The very limited amount of time available to them means that many working people have adopted some unhealthy aspects of western lifestyles. For example, instead of having a proper meal in the morning, many people choose tea and toast because they can be prepared quickly. In order to avoid heavy traffic in the morning and evening, some people routinely have breakfast in the office cafeteria and dinner in a restaurant. They only have meals at home on the weekends.

Despite the problems imposed by driving in Bangkok's traffic, most residents no longer see owning as car as a luxury, but as a necessity – important to avoid the slow and crowded buses and severe air pollution. Also for those living at the fringes of the city, where there is no public transport, a car is essential. Most residents that drive their own cars feel that there is little that can be done to solve Bangkok's traffic woes, and so are resigned to the situation. Many people with cars did what they could to relax while caught in the traffic: reading magazines, listening to music or meditating.

People with cars ensure that they suffer the impacts as little as possible. However, by buying a car then cocooning themselves in air-conditioned comfort, they add to the traffic-related problems that they are trying to escape. Few people that currently drive cars would be prepared to use buses, even if bus services were improved and traffic jams reduced.

> Thai people have certain expectations [driving a car is more prestigious than using a bus] .... I don't think Thai people will put on a suit and necktie then go to work by bus.

Conflicts in Managing Bangkok's Transport System

Solving Bangkok's traffic problems involves managing a large number of interrelated issues. Any changes in the transportation system will affect not only all ten million people that live in or commute to Bangkok every working day, but also thousands of organisations – businesses, factories, government departments – that have a stake in or responsibility for improving Bangkok's traffic. In this complex situation, wherever individuals or groups come together, conflict is likely. Changes in the structure of the Thai government, economy and society generally are also a source of conflict, and this has to be factored into any proposed solutions.

While Bangkok's residents fight one another on the city's roads, other battles take place between those responsible for the management of Bangkok's roads and transport system.

The Traffic Police

Caught literally in the middle of Bangkok's traffic nightmare are the traffic police. They have to deal with not only frustrated motorists, but they have to battle government departments intent on other forms of traffic management (such as the construction of expressways and flyovers).

The head of traffic police complained in February 1992 that his department could not improve matters further while the government policy of constructing new freeways encouraged more car drivers onto congested roads, rather than using public transport.

Government Authorities

Government departments compete with each other. Traditionally they have worked in isolation from one another, with little communication or coordination between them. Although the newly established system of interdepartmental

161

coordinating committees improves communication amongst government agencies, it also opens up new arenas for conflict and competition.

The lack of coordination between authorities, and the poor division of responsibilities between departments results in chaos on the city's streets. For instances, roads repaired by workers from one department one month (at considerable disruption to traffic) may be torn up the next by workers from another department installing local power or water supplies. Government departments are loath to give up projects or responsibilities even in the interests of rationalisation.

Another source of conflict for government departments are the differing interpretations of the policy of modernisation. Infrastructure programmes, such as urban redevelopment and roadworks, have to compete with economic development programmes sponsored by powerful organisations such as the NESDB for their funding.

Despite the fact that many government employees would agree that greater openness would result in better and less complex government, the change will be painful for many. For instance, the adoption of a merit-based promotion system is slowly replacing the familiar client–patron relationship of advancement. This imposes a new and unfamiliar set of rules of conduct and behaviour on public servants, which for some involves a loss of security and protection.

Society

There is an increasing involvement in government and politics by ordinary citizens and non-elected organisations (as will be discussed in greater detail later). This poses a great challenge to both politicians and public servants that have previously not expected involvement (or 'interference') from the community. This has resulted in considerable friction, which is certain to increase in the future as the community becomes more vocal and expectant of government.

WHAT IS BEING DONE TO CHANGE THE SITUATION

Political

For politicians, changes in Thai society used to mean redefining whom they serve and respond to. Under the traditional hierarchical system, the public had no input into political decision making. Corruption was rife as politicians responded to power groups such as business, family or social interests, rather than their constituencies. Today, politicians are expected to be honest and accountable.

In the political arena, there is continuing tension between democratic forces and the military. The military has dominated Thai politics since 1932, and will be loath to surrender its position permanently. On the other hand, ordinary Thais have increasing expectations of government, such as accountability and representation. The conflict has cost lives on three occasions in the past twenty years.

An important change in Bangkok's political landscape came in 1985, with the first election of a Governor by the people of Bangkok. Previously, the Governor had been appointed by the central government as Governors in the provinces still are. Elected leadership is producing noticeable, if still relatively minor, urban reforms and encouraging public expectations of responsible government. The Bangkok Metropolitan Administration (BMA), the government agency with responsibility for Bangkok, is responding to this leadership. The BMA and its political leaders are emphasising Bangkok's urban problems, including traffic, as a matter of priority, and have ordered the implementation of some measures. They are hampered in their efforts because the BMA has only limited powers, and they have to contest with national government departments that have jurisdiction over some local government activities.

The local political party currently in power in Bangkok has made a virtue of honest politics – a reaction against the corruption (considered excessive even by Thai standards) that persisted up until the early 1990s. Corrupt practices, such as vote buying and bribery, are now liable to shameful public exposure. The party has proved responsive to its electorate, and is educating city residents to have certain expectations of the political arm of government. This move can be seen against a background of increasing expectations of politicians by Thais at all levels, as witnessed in the mass protests against the appointment of a military Prime Minister in 1992.

At the national level, increasing numbers of politicians have become aware of the problems of urban development in Bangkok, and the results for the transport system. Politicians with Bangkok electorates have been active in pushing reforms – both in government policy and practices.

Bureaucracy

The Thai government and public service combines elements of traditional Thai culture and contemporary western government.

At present, most public servants tend to spend their whole career in one department, which limits their experience and contacts. Public service careers are still influenced by cultural traits such as patron-client relationships and traditional hierarchical/hereditary relationships. Recently, however, there has been some encouragement of training and the merit principle.

The merit system is having an effect in senior public servant advancement, and encouragement of university training is paying off in two ways. Educational levels are increasing, and there is more communication across departmental boundaries following lines of university friendships. There is also an increasing interaction between academics and policymakers.

However, the public service and government departments are hampered in implementing new government policies because of poor structures. Division of responsibility is somewhat arbitrary, and there is no clear distinction between the national and local government departments with responsibility for Bangkok.

There are some changes being made by the government to deal with the traffic and some of the problems it creates, such as air pollution. The government has begun a campaign to phase out leaded petrol in an attempt to reduce environmental lead levels and the incidence of lead poisoning. The government has also introduced tough new standards for noise pollution and exhaust emissions for cars, buses and motorcycles. Emission levels for new cars have been set at levels accepted by Japan, the US and European countries. All diesel and motorcycle engines are required to be fitted with 'particulate control technology' to reduce SPM levels. Other solutions being explored by various government departments include the development of an electric rail system, extensive bicycle ways and the introduction of a 'quiet zone' in the city centre, where cars will be banned except on a permit basis.

One factor hampering all of the plans and reforms is the isolation in which all have been developed. It is inevitable that they will be hard to integrate into a workable system.

The Traffic Police

The traffic police have changed rapidly as an organisation in recent years. The widespread bribery that once flourished amongst the ranks has been largely eliminated. The police have taken a different approach to many government authorities: instead of relying on engineering 'fixes' to ease Bangkok's traffic problems, they have begun to move to an integrated set of social and technical solutions.

In trying to improve traffic flows, the police have aimed at changing the behaviour of drivers – emphasising traditional Thai values of tolerance and courtesy. In this they have used the media to create a public spirit favouring considerate driver behaviour such as keeping to lanes, and to encourage other drivers to use social sanctions such as horn-blowing against transgressors. At the same time, they have stepped up enforcement. The police have also removed roadside and pavement obstructions such as hawker's stalls (despite some controversy). They are in the process of installing an electronic coordination system for traffic lights. Their efforts have meant that traffic flows are now more orderly, if not actually faster, than a few years ago.

Business

Business and industry are not unaware of the problems caused by the traffic and poor urban development. Many suffer loss of production as a direct result of the traffic. In recent years, many businesses have begun to lobby the government for improvements in Bangkok's infrastructure.

The motivation for change has come from several directions, not solely with a loss of productivity and profitability. As noted earlier, Thai business cannot be solely characterised as profit-motivated, and some private companies and individuals such as the media owners, have been active in promoting social change.

Another motivation has been the realisation that stability in society and particularly government is beneficial to business, investment and trade. Stable government implies stable legislation and regulations which make planning long-term business easier. The eradication of corruption and vote-buying has also led to improvements in long-term business prospects. These realisations have contributed to an increasing push amongst the business community for greater democratic and social reform.

Community and Non-government Organisations

Within the community, there are a number of influential reform-minded individuals, many of them senior Buddhists and intellectuals (intellectuals, academics and students play a significantly greater part in public life than their counterparts in the West, many of them seeing participation as a social and moral responsibility).

Popular participation in government is increasing. Non-government organisations have been involved in much of Bangkok's urban renewal, such as improvements in drainage, electricity and roads in slum areas. Increasing involvement by the community in government could be seen as a threat by some in the government bureaucracy, as it is a significant change to the traditional Thai hierarchical society in which the public had, and expected, no status to influence political or bureaucratic decision-making. In fact, there is a growing climate of acceptance for public participation, provided that it is able to transcend sectional self-interests.

Thai community groups and non-government organisations have proved capable of developing novel and effective solutions to other social problems in the past. A good example is the family planning programme of the 1970s, which appealed to the Thai sense of *sanuk* (fun) to create a climate for public discussion and education about family planning. Police handed out condoms on the streets, and there were many witty publicity stunts. Condoms are now known as 'Meechai' after the chair of the Family Planning Association. A restaurant, *Cabbages and Condoms*, owned by the Association, continues the fun theme.

A novel and popular method of coping with Bangkok's traffic was the introduction in 1990 of an FM radio station devoted to monitoring traffic flows through helicopter observation, and through reports phoned from commuters (on their car phones) and residents. The station was originally set up by a private company in cooperation with several government departments. Five years after its inception it has mobilised a large number of volunteer reporters throughout the city and suburbs. The station now broadcasts eighteen hours a day, and enjoys a very enthusiastic level of participation, as the availability of information and the talk-back sessions provides a release for travellers' stresses. Local 'FM100 Clubs' are engendering a new sense of community. It is listened to by virtually every road user in Bangkok before and during their trips for news about accidents and traffic jams.

Among the community, there is increasing confidence in 'home-grown' solutions to Thai problems, rather than reliance on western models for solutions. For instance, the reopening of several *klongs* in recent years, and attempts to manage the pollution in existing waterways to make them acceptable for, among other things, transport is an example of how traditional Thai methods are being adapted to modern life.

Prospects for partnership between the community and government are now improving. Leading individuals from each stakeholder group came together in January 1994 to form a group called Traffic Crisis '94. About 100 NGOs are expected to join. The group's committee has undertaken to work out a master plan for Bangkok's traffic, including short-, medium- and long-term solutions. Small workshops are examining facets of the transportation system, including mass transit systems, expressways, cargo transport, bicycle lanes, limitation of private cars, land use, city planning, minor road networks, regulations, and traffic management for big buildings and public places. The group includes leading social critics, a popular former Prime Minister, academics, city planners, engineers, bankers and hoteliers. The initiative has come from the non-government sector, but has broad government support. Stakeholders are at last producing ideas together: can appropriate action follow?

References and Further Reading

Abt Associates and Sobotka and Co (1990) *Ranking environmental health risks in Bangkok* US Office of Housing and Urban Programs, Washington

Bhongmakapat, T (1990) Prospects and strategies for Thailand in the 1990s Proceedings of the symposium on *Directions for Thailand reforms in the 1990s* Faculty of Economics, Chulalongkorn University, Bangkok (in Thai)

Boyden, S, Dovers, S, Shirlow, M (1990) *Our biosphere under threat* Oxford University Press, Melbourne

Department of Land Transport (1990) *A survey of hired motorcycles in Bangkok between 18 January–24 February 1988* Unpublished report of the Transport Statistic Section, Department of Land Transport, Bangkok (in Thai)

Dhiravegin, L (1983) Bangkok-centralism: Roots and Problems Proceedings of the annual symposium on *Thailand is Bangkok?* Faculty of Economics, Thammasat University, Bangkok (in Thai)

Division of Health Statistics, Ministry of Public Health (1989) *Public Health Statistics*, 1987

Hutaserani, S and Jitsuchon, S (1988) 'Thailand's income distribution and poverty profile and their current situations' Paper presented at 1988 TDRI year-end conference on *Income distribution and long-term development*

Ingram, T (1955) *Economic change in Thailand since 1850* Stanford University Press, Stanford, California

Japan International Cooperation Agency (1989) *Medium to long-term road improvement/management of roads and transport in Bangkok*

Kangagate (forthcoming) Measures to be published in the integrative report of the research project *Impacts of modernisation and urbanisation of Bangkok*, (H Ross et al (eds))

Karnchanapant, S (1989) 'Development alternatives: The case of green-belt in Thailand' in Faculty of Arts Silpakorn University Journal, vol 12 p 1 (in Thai)

Meaksupa, O et al (1987) *Presentation of psychological problems among Bangkok's population*. A Survey report (in Thai)

Phipatseritham, K (1983) Thailand is not Bangkok Proceedings of the annual symposium on *Thailand is Bangkok?* Faculty of Economics, Thammasat University, Bangkok (in Thai)

Police Department (1990) *Crime status in Thailand, 1988* Research and Planning Division, Bangkok (in Thai)

Poungsomlee, A (1991) *An integrative study of an urban ecosystem: the case study of Bangkok* PhD thesis, Australian National University, Canberra

Poungsomlee, A and Ross, H (1992) *Impacts of modernisation and urbanisation in Bangkok: an integrative ecological and biosocial study* Institute of Population and Social Research, Mahidol University, Bangkok

Ross, H *Opportunities for change* forthcoming in the interprative report of the research project Impacts of Modernisation and Urbanisation in Bangkok, H Ross et al (eds)

Suwarnaret, K et al (1989)(unpublished) *Environmental problems due to urban transport in Bangkok*

Taweesuk, S (1990) The need for green areas for people's recreation in Bangkok Proceedings of the first joint annual conference on *Quality of life of city people* Mahidol and Thammasat Universities, 19–21 March 1990, Salaya, Nakornpathom (in Thai)

Thailand Development Research Institute (1990) *Energy and Environment: Choosing the right mix* Research Report No 7, TDRI, Bangkok

Thailand Development Research Institute (1993) *National development framework: final report* TDRI, Bangkok

United Nations (1987) *Population growth and policies in mega-cities: Bangkok* Population Policy Paper No 10, Department of International Economic and Social Affairs, UN, New York

Wyatt, D (1984) *Thailand: A Short History* Yale University Press, New Haven and London

9   The Future of Calico Creek

Local management of a small subtropical water catchment

D Ingle Smith, Linden Orr and Rob Wiseman

Introduction

Calico Creek is a small agricultural community located on the catchment of the Sally River in southern Queensland. The area is subtropical and lies amid the steep hills at the edge of the Great Dividing Range, 80 km inland from the Queensland coast. Calico Creek is 10 km west of the township of Gympie (population 50 000) and comes under the jurisdiction of the local elected Shire Council.

The headwaters of Calico Creek arise in what locals call the 'slip country' – a reference to the hilly region prone to landslides in the wet season. Local landholders recall with delight the time a banana plantation slid into a neighbouring property. Another season, and another farmer awoke to find he had acquired a new dam, courtesy of the slide of a neighbour's property.

The area, which was once covered by subtropical eucalypt forest, now supports a small farming community whose income is derived predominantly from agriculture. Produce includes small crops such as beans, squash, zucchinis, tomatoes and cucumbers; tree and vine crops such as paw-paw, avocados and passionfruit; and dairy cattle and a few beef cattle. Patches of the original forest cover remain – although mostly restricted to the steeper hills and gullies. The Calico Creek catchment adjoins a small State Forest reserve.

The water in the creek is unpolluted and untreated, and the supply is usually abundant. Small dams and weirs dot the countryside, and together with bores, they supplement water supplies from the creek. Most of the water taken from the creek is used for irrigation of crops and pasture, with watering livestock (mostly dairying) and domestic supply making up the remainder.

By Australian standards, the region is well watered, receiving approximately 1000 millimetres (40 inches) of rain in an average year. Like the rest of the country, however, rainfall and runoff are both subject to significant variation. Floods and droughts have affected the region in the past. Severe water shortages are a feature of rural life. The region suffered particularly in the drought of 1982–83 and again in 1991. In periods of low rainfall the flow in Calico Creek can drop to nothing, and water quality declines dramatically. In the 1982–83 'dry', the mineral content (total dissolved solids) was too high for domestic use. Residents were forced to rely on storage from roof catchments.

About fifteen years ago, an irrigation scheme based on a new dam was established to support the neighbouring catchment of Meatloaf Creek. The Commissioner for Water Resources of the time was not in favour of the proposal, but political pressure was brought to bear, and the scheme was developed at a cost of $1.4 million. The Commissioner's view has been vindicated because at no stage has the scheme ever paid for itself.

Pressure from the Calico Creek community resulted in the construction of a pumping station to deliver water from Meatloaf Creek to the lower part of Calico Creek. Farmers on this section of the Meatloaf diversion, downstream of Calico Creek, now have access to a regulated water supply – and pay accordingly. Users are charged $39.70 per megalitre for a specific allocation of water with 50 per cent of this fee to be paid whether the water allocation is used or not. One farmer stated 'It's expensive to be part of the scheme, but it's worth holding the licence for the dry times … it's our drought insurance'.

Users of the unregulated supply in Upper Calico Creek do not pay a fee for the water they abstract; the problem is how to allocate the creek water at periods of low flow. There are approximately twenty landholders above the regulated supply with blocks ranging in size from 5 to 340 acres, whose farms are run essentially in family units. This case study is concerned with this community.

Calico Creek illustrates the problems of sharing a finite resource among a number of users. Water is a renewable resource but the marked variations between seasons and between years cause conflict between users with differing needs and perceptions. Increases in the number of users leads to greater scarcity at times of decreased water availability. Supply is constant in the long term but variable in the short term, demand is consistently upward and increases at times of low stream flow when demand to augment rainfall is at its peak.

The number of farms and households that abstract water from the unregulated section of Calico Creek is small. However, the diversity in size, water use and the problem of adding 'new' players combine to make the problem of managing the resource complex. The possibility of change, ie from unregulated flow to a dammed regulated catchment, poses an additional choice that creates a further series of problems.

An attraction of Calico Creek, as an example of conflict resolution in resource management, is that the community has essentially managed water allocation for the past forty years. Although superficially simple the case study illustrates the complexity of choice. The problem is a contemporary example of the problems of managing 'the commons'.

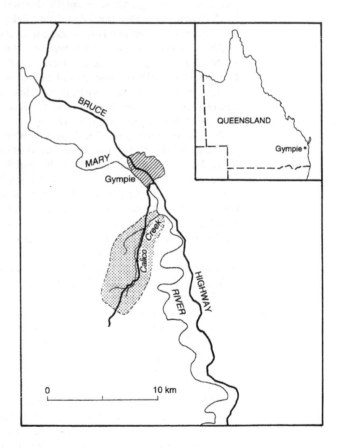

Figure 9.1   Calico Creek: a small sub-tropical water catchment area

WHY CONFLICT MANAGEMENT FOR CALICO CREEK?

Conflict in Calico Creek arises primarily through the allocation of water for agricultural purposes, what farmers and irrigators are prepared to do to get the water they need, and how others react to their actions.

Calico Creek is divided into two sections – unregulated and regulated. Without the security of large-scale water storage facilities, unregulated catchments are particularly susceptible to drought conditions. Regulated catchments have access to a dam which stores water for controlled releases during times of low flow. Irrigators on unregulated catchments do not pay charges for the water they use – those on regulated catchments do. In both cases there are controls on irrigators regarding the quantity of water that can be abstracted from a creek or river during dry periods. Restrictions can be in the form of limiting the number of hours water may be pumped, or limits on the volume of water pumped. Although irrigators on the unregulated section of Calico Creek do not pay for the water they use, they have to hold a licence to abstract.

There are two problems with river supply regardless of the presence or absence of a dam/irrigation scheme. These are:

1   allocation of licences to abstract, who is eligible and how much water they can use;
2   the imposition of restrictions on water use at the time of low flow or drought.

Water Supply

The major management question and source of conflict in Calico Creek is how should the supply of water from Calico Creek be shared, especially in times of low flow. Legally, the State water agency has control over the allocation of water, but it has developed a system to allow for community input – the Water Advisory Committee (WAC). There are 50 WACs in Queensland, the first of which were established in the 1950s. WACs are locally based and typically each have 20–200 members, representing the water users within a catchment. Their role has been to 'assist the Water Resources Commission by advising when pumping for irrigation shall be allowed on streams in times of reduced flow and to offer comments on new licence applications for their particular area'. Although the WACs are purely advisory and have no legislative basis, it has been the Commission's policy to follow WAC advice, even when it is contrary to the Commission's view, either locally or State-wide. One former Minister warned in a pencilled minute, 'we go against the advice of a WAC at our peril.'

The Calico Creek WAC comprises ten of the twenty landholders licensed to pump from Calico Creek, all of whom are situated upstream of the portion of Calico Creek with regulated supply. The other landholders are not involved either because they use little water directly from the creek or have access to the irrigation scheme downstream. A representative of the district Water Resources Commission always attends the WAC meetings to provide technical advice and explain the Commission's views.

In periods of low flow, the WAC may impose restrictions. These may be limitations on the time for which pumping can be allowed (perhaps modified to allow for pump size), or limitations on the volume of water that can be pumped and when it can be pumped.

The WAC may also discuss issues such as applications for new pumping licences and proposed land subdivisions on the Creek. Anything a farmer does to a creek could affect the downstream water supply and is therefore of concern to the WAC.

In the past, the WAC has resolved conflicts by calling a meeting and discussing the problem. The irrigators may come to an agreement through talks, or the issue may be put to the vote. Usually discussion is sufficient for all parties to come to a unanimous decision – although sometimes a vote may be necessary. If agreement still isn't reached, the Water Resources Commission may act. Decisions are made 'just by mutual agreement .... If we can't reach agreement, then the WRC bloke measures the stream flow and divides it up.'

Meetings are normally called when one of the WAC members has something he or she feels needs discussion. Frequently, the problem is that one of the irrigators at the downstream end of the creek has insufficient water to permit pumping. If someone does not abide by WAC decisions – usually water restrictions – they can be 'gazetted' prior to legal proceedings. Peer pressure and the threat of public embarrassment is usually sufficient to enforce sanctions.

Peer pressure is one of the most effective weapons in the WAC armoury because everyone wants to get on well with their neighbours. This is a double-edged sword, however – it means that people can be bulldozed into voting for issues they don't want to support: 'Sometimes we just have to go along with it' is the view of one member, who often finds himself in the minority. It also means that landholders are reluctant to become involved in court cases (even ignoring the cost which they would have to bear). Someone more immune to this type of pressure, such as a large or absent landholder, is less bound by the WAC or their neighbours. For instance, one landholder discussed the amount of water each person should receive:

> One person above us takes all the water. I rang him a week or so
> ago to find out his pumping time so that I could work around his

times. But the other day, his pump was going at a time not agreed – we don't really know what time he's pumping. I went down to the creek and we had no water. In dry times, if he is pumping, then there's not much water left for us downstream. If he's pumping, it takes a few hours to get a flow again.

Personalities play an important part in the operation of Calico Creek. A change in the Chair of the WAC can have profound effects upon the way that body operates. Most of the WAC's sanctions are issued on trust. As one farmer said, 'If a man is not honest about his pump, we know what he's doing. We're not silly. We can't do much more than that though.'

There are tensions within members of the WAC. Some irrigators feel that one farmer would pump all the water he could from the creek if he could get away with it. This farmer happens to have the largest property in Calico Creek, the largest number of licences and is the Chair of the WAC. Some landholders see themselves as powerless to change the situation.

There is also tension between the farmers and the Water Resources Commission. For instance, local farmers on the regulated section of Calico Creek called 'unfair!' when they had to pay excess water rates during times of high flow and when the Meatloaf Dam is full. Another issue is the provision of water rights: the Commission is aiming to provide the community equitably with the water that it needs, while irrigators with riparian rights (direct access to the creek) feel that they have a priority over non-riparian landholders in water use as they paid higher prices for their blocks. There is also tension between landholders on this same issue:

> ... we used to get annoyed about the dry farms who paid less for their property but they can pump from the creek. In dry weather it's not fair. We bought the property for the creek.

Despite these differences, the Commission has to rely on irrigators and the WAC to police its regulations. Neighbouring irrigators '... keep a good check on you by driving up and down the road to see if you're pumping.'

Other Sources of Water

There are three additional options to water users in Calico Creek that enable them to obtain water other than from the creek itself. These are:

1   roof catchment
2   farm dams
3   groundwater pumping.

Roof catchment has a limited potential and is normally used in Australia to augment domestic water supply. With properly managed rainwater tanks, the water quality is often superior to other forms of supply. If water is also collected from roofs of outbuildings it can come close to meeting household needs.

Small farm dams can be constructed on minor gullies and tributaries. They can augment supplies from the main creek. This form of water collection is especially useful for stock and can be regarded as a private source of supply. Small dams can be constructed without the need for State approval and licensing. They are used by some farmers in Calico Creek.

The major alternative source of water supply for irrigation is from **groundwater bores**. In practice, bores of this kind can be installed in the Calico Creek region without Commission approval or licensing. Clearly, the discharge from such bores depends upon the local geology and how good the aquifer is. The Calico Creek region as a whole has available renewable groundwater resources that are similar to the average divertible surface resources. Problems can occur when groundwater bores and pumps are installed close to surface streams. In such cases, pumping diverts water from the stream – this would appear to occur in Calico Creek.

Ideally, groundwater provides a back-up source of water that can be used at times of low flow. The barriers are the costs of installation of the pumps and the availability of suitable aquifers on the farm property.

There are two other issues of concern to those in Calico Creek: subdivision of land and market prices for small crops.

Subdivision of Land

There is increasing demand for residential and small farm blocks from city dwellers. Some sectors of the Shire Council would like to see this demand met, as it would raise extra revenue. Most farmers in the Calico Creek catchment are opposed to this type of development. They see it taking up prime agricultural land, as well as eating into their water rights.

> People with jobs in the city want water that we really need. .... We've recommended that they don't have licences during restrictions ....
> Who do we look after? We have to look after agriculture ....

This statement also gives some idea of the acceptance of stewardship of the land by long-term families and distrust toward those from outside the area – especially those from an urban background.

Market Prices for Small Crops

Prices for beans and other small crops, a major source of local income, have been falling for some time. Many of the farmers in the area are losing money and are under pressure from the banks. One farmer reported 'I've been in beans for 16 years and I've never seen it as bad as this. I'm glad I got out of it two years ago'. Another said, 'What is the point of having our water if we have to pay to grow our crops'. Currently, the market price for beans fluctuates around 90 cents per kilo and it costs the farmer $1.00 per kilo to grow, pick and pack the beans. The produce from Calico Creek is taken to a local wholesale market and then transported interstate for processing.

WHAT IS THERE TO MANAGE IN CALICO CREEK?

Upper Mary River Catchment – Water Resource and Use

This section explains the larger regional catchment of the Mary River in which Calico Creek is situated (see Figure 9.1). It provides a guide to the regional water resource and use for the year 1985. It also provides a background against which to judge future demand. Similar data are available for the whole of Australia and are given in the *1985 Review of Australia's Water Resources and Water Use* (DPIE, 1987). Proportions of water usage can be assumed to be reasonably close to those for Calico Creek.

The Resource

The Mary River catchment has an area of $14\,600\,km^2$ and a mean annual runoff of $4110 \times 10^3 ML$, the divertible surface supply – the long-term average volume that can be used on a sustainable basis – is estimated to be $650 \times 10^3 ML$. Groundwater reserves have the potential to supply a further $640 \times 10^3 ML$.

The combined divertible, surface and groundwater resource is $1290 \times 10^3 ML$. The surface water is of high quality with no use restrictions in terms of salinity. Locally, the groundwater has limited salinity problems but these are not considered to present major problems for rural use.

Current Use

Only $67.0 \times 10^3 ML$ were utilised in 1985, and the figures do not appear to have changed significantly since then. This is just over 5 per cent of the available water. Details of individual uses are given in Table 9.1. The population for the

175

Upper Mary catchment is 158 000 of which 35 000 is classified as rural.

Table 9.1 Water use (10^3 ML) in the Upper Mary Catchment

Urban and industrial

	Reticulated	Self-extracted
Domestic	20.24	-
Industrial	1.21	3.93
Commercial	1.41	0.25
Sub-total	**22.86**	**4.18**

92.5% of these supplies are from surface water sources.

Irrigation

Pasture	12.39	(4170 hectares)
Crops	14.36	(4260 hectares)
Horticulture	11.12	(3710 hectares)
Sub-total	**37.87**	

84.5% of irrigation water is from surface water sources.

Rural

Stock	0.80
Domestic (and other)	1.17
Sub-total	**1.97**

All water for rural use is from surface-water sources.

Note: Overall the urban and industrial sector accounts for 40.4% of total water use, irrigation for 56.6% and other rural uses for 3.0%.

Irrigation, from either regulated or unregulated sources, does not provide water security for agriculture. The concept of irrigation providing a technique of drought-proofing agriculture is an illusion. It is essential that drought is factored into normal farming practice. This is the major finding of the National Drought Task Force which reported in 1989, who recommended that Federal contributions for drought relief cease in mid-1991.

There is no doubt that the construction of a headwater dam in Calico Creek would improve the reliability of supply at times of low rainfall. However, this would result in charges for water used. It is now policy that such charges should match the supply costs, including a major contribution to the construction of the dam. This contrasts with the former situation in Queensland and elsewhere, in which the construction costs were borne by the State. This move to 'user pays' is now the accepted goal for water supply throughout Australia. It matches comparable trends in the supply of other services.

The needs vary for differing crops. Small but continuing supplies are required for livestock, however long the drought. This contrasts with vegetable growers who need to sacrifice some or all of their crop when supplies are partly reduced. There are examples of members of Calico Creek's WAC, who are vegetable growers, recommending that irrigation pumping cease in order to allow dairy farmers to water their stock for a longer period.

It is possible for individuals to outlay capital to provide additional farm dams or bores on their own properties. The future scenario includes the provision of additional supplies for residential uses often located some distance away at coastal resorts. Primary producers find it hard to agree to the use of residential supplies to water unproductive gardens in times of drought. The alternative is for water to be trucked in: this is very expensive.

Stakeholders

- The landholder with the largest property:
  - has four licences to pump from the Creek;
  - is the chair of the WAC;
  - has the most dairy cattle in the district.
- The other landholders, both riparian and non-riparian:
  - riparian landholders have paid more for their land so they believe they have priority on water use.
- The Stream Control Officer from Water Resources Commission:
  - wants the farmers to make the decisions but realises there will be conflict.
- The Local Shire Council:
  - wants subdivision to occur in the area as increased rates mean more revenue.
- The Land Court:
  - has been involved in the past in litigation between farmers in the area.
- The local Member of Parliament:
  - is friends with the largest landholder on Calico Creek.
- The produce market wholesalers:
  - have a vested interest in produce from the Calico Creek area as there is a cash economy operating at the markets.

Services/agencies

There are several other organisations with interests in or responsibility for the aspects of Calico Creek.

- The State Department of Primary Industry:
  - revenue from produce
  - quality and quantity of produce
  - farm management
  - economic advisory service
  - agronomist advisory service
  - statutory marketing boards
- The Water Resources Commission:
  - revenue from licences
  - water usage
  - water quality and supply
  - catchment management
  - pumps, dams and bores
  - sand and gravel extraction
  - creek excavation
- State Department of Environment and Heritage:
  - pollution control
  - water quality
  - land degradation
  - salinity
  - conservation
- State Health Department:
  - quality control of meat and milk
  - public health
  - family and personal health
- The sitting Member of Parliament and his political party:
  - votes
- Didgereedoo Shire Council:
  - land use
  - revenue from rates
  - subdivisions
  - boundaries of properties
  - pollution
  - local economy
  - water quality and supply

HOW CONFLICTS ARE MANAGED IN CALICO CREEK

Conflicts regarding the use and allocation of water resources in Calico Creek are normally handled by the WAC. By and large it is effective in imposing and policing sanctions. Peer pressure and the threat of public embarrassment are usually sufficient to keep it running effectively. The community benefits because the allocation is determined by those directly concerned, while removing the problem of imposing unpopular and possibly inappropriate decisions.

The WAC is not perfect however. Many of the farmers do take more than their allocation – some more publicly than others. Some jam the water meters with sticks so that the amount of water pumped is greater than the amount recorded. Others pump for longer periods than permitted, or at different times. One farmer has a bore beside the creek, so that although technically he is pumping groundwater (and does not have to meter the amount of water he abstracts) in reality he is pumping from the creek. However, it is very hard to substantiate how much river water is being pumped. Some irrigators have bragged about the amount they have pumped, although most were more circumspect: 'You get your fingers burnt if you talk too much about it.'

The problem with giving the power over water restrictions to those who need the water is that it may lead to a reluctance to impose restrictions until it is too late for them to be of value. How representative the WAC really is, and what conflicts and disruption it may cause amongst member farmers are two other issues that need attention.

The WAC also has to work within the State legislation. For instance, livestock has priority of water use over irrigation and pasture crops. The State Water Authority does have the power to impose water restrictions that are legally enforceable – although it readily admits that such controls are difficult to police. In the event of a breakdown in the operation of the WAC, the Water Authority can and will intervene.

By and large, however, the WAC is a good balance between the potential scenarios of:

- **a free-for-all** – where irrigators pump as much as they want, regardless of water flow and the needs of those downstream; and
- **State imposed restrictions** – which are generally imposed on a State-wide or regional basis, not at the scale of Calico Creek. State imposed restrictions are not responsive to rapidly changing situations, and are extremely difficult to police, especially in small or remote areas.

If the WAC is not able to solve a problem, there is the option of prosecution through the Land or Civil courts. One major problem with prosecution, especially in the Land Court, is that it can take up to twelve months to arrive at a verdict. There is also the necessity of having hard evidence of wrong-doing or illegality. One Calico Creek farmer has a series of interconnected bores and dams to supplement what he pumps from the creek, and so it is difficult to say exactly how much water he is abstracting. Some of his neighbours would like to see the Water Resources Commission prosecute him, but realise the difficulties of doing so without proof. Farmers are reluctant to resort to the courts to resolve conflicts. Apart from cost and time, most farmers do not want to incur the disfavour of their neighbours.

But the WAC may become a mechanism to support the *status quo*. As one local put it, the WAC system '… has worked for years. If a new person came into the district and wanted to change things, it would be difficult.' As the WAC recommends on the issue of licences, it effectively has the power to block change. The WAC is not dedicated to broad-scale environmental policy but in protecting the interests of its members. As one WAC member said bluntly, 'We look after our own interests'. Another said that despite restrictions, a WAC meeting decided that 'if there was water going past J--'s pump, then we could increase the amount of water we could pump'.

The issues facing Calico Creek cannot be solved easily. How to proceed is a difficult question – should the WAC be maintained as an effective method of managing Calico Creek, or should it be replaced, and if so, with what? How are the water management issues facing the area to be dealt with? What can be learnt from past experience?

References and Further Reading

Department of Primary Industry and Energy (1985) *Review of Australia's Water Resources and Water Use* Australian Water Resources Comment, Australian Government Printing Service, Canberra, 2 vols

Part 3

Managing Conflict and Change:
A Learning Programme

10 Introduction: Setting the Agenda

Identifying the options

LEARNING MATERIALS

- Negotiation
- Conflict resolution styles
- Tragedy of the commons
- Brainstorming

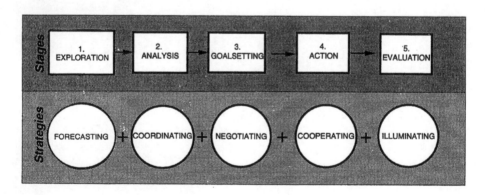

Figure 10.1 Identifying the options

AIMS AND OBJECTIVES

During the introductory session, participants will be able to:

- identify the five principal stages of environmental management;
- review the range of conflict management strategies available to environmental managers;
- share their objectives and resources;
- recognise the risks and opportunities in practising environmental conflict management; and
- identify their own conflict management styles.

INTRODUCTORY SESSION

The first task in approaching environmental conflict management is to determine *why* there is concern about environmental conflict management, *what* is to be managed, *who* is to do the managing, and *how* the management is to be conducted. Chapters 2-5, 'Why, What, How and Who', review each of these issues in turn. Each participant should read that section before starting the course. During the introductory session, participants will share their individual perspectives and experiences on these aspects of environmental conflict management.

About a week before Session 1, request descriptions of each participant's aims and objectives for the course, and the personal experience which they bring to the course. These aims and experiences are shared between participants during the introductory session. There are opportunities for each participant to identify the conflict management styles where they apply in different situations, and an exercise demonstrating the 'tragedy of the commons' for environmental management. (See The Developers of Id, page 196.)

Each learning module is organised in the same way:

- a description of the aims and objectives of the session;
- a review of the risks and opportunities offered by conflict management at each of the five stages of environmental management in turn;
- a set of activities which provide experience in the conflict management skills likely to be most useful at each stage;
- readings which may be used before and after the activities; and
- a set of materials for the session's activities.

RISKS AND OPPORTUNITIES

The theme common to each of the learning modules is that conflict is the inevitable accompaniment of change. The challenge is therefore not to prevent conflict arising, but to identify the outcome of the conflict and the best ways to manage it.

A mountain symbolises the issues faced by anyone managing environmental change. At the summit of the mountain is:

- cooperative teamwork, with the goal of achieving a synergy of solutions of mutual advantage to all interests.

At the base of the mountain from where any climb has to begin are:

- isolation, the decision not to engage in the debate at all; and
- confrontation, in which positions have been adopted in fixed opposition to one another.

These three positions can be adopted equally well by the participants within the course processes as well as by interest groups involved in managing environmental change. The course is designed to reduce the risk of isolation and differences due to confrontation between participants, and to maximise the opportunities for balanced, cooperative teamwork.

Whichever way the programme is managed, there are some basic essentials for its success:

- All participants (including the course coordinators) should agree to share responsibility for the outcomes of each exercise. There are no winners or losers in the modules, only learners.
- All exercises are simulations of potentially real events. The participants will need to spend time sharing their experiences after each exercise, both to maximise the learning and to ensure that no negative impressions are left.
- Each exercise and each participant is regarded as making a valuable contribution to the group's professional experience.

ACTIVITIES

Theme: Relationship Between Environmental and Conflict Management

The learning programme itself presents risks and opportunities to both course convener and all participants. The Course Coordinator's Guide (Appendix 1)

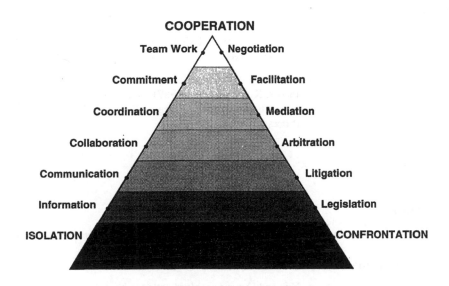

Figure 10.2 Managing environmental change

contains the groundrules for designing and presenting the whole programme, and the materials and guidelines for exercises for each session. The Course Coordinator's Guide can equally well be used for individual learning, or by participants who wish to design and implement their own course.

The Conflict Management Mountain is described in some detail in Chapter 4. Organise the following activities as preparation for the programme:

1 **Seminars:** short papers on resources management, the potential conflicts, the elements of conflict management (60 mins).

2 **Introductions:** participants describe their aims and objectives for the course and their own experiences with environmental management (30 mins).

3 **Conflict resolution styles:** a diagnostic test of five styles of conflict management is discussed in relation to participants' experience. Materials provided below (30 mins).

4 **'The Developers of Id':** a recreation of the 'tragedy of the commons'. Materials provided below (60 mins with discussion).

READING

The learning programme is designed to be internally complete with readings from the following texts:

- Chapters 1–9 of this volume.
- Fisher, R and Ury, W (1987) *Getting to Yes* Arrow, New York, for elements of constructive negotiation, called the Harvard Negotiation Process.

MATERIALS

Negotiation

Ian McAuley, Consultant, 24 Schlich Street, Yarralumla, Canberra, Australia

The General Approach

- You and I have a problem, let's see how we can get together and solve it (negotiation as problem solving).
- Let's not rush into a solution; we have certain interests to attend to. What is the most productive way we can attend to our interests?
- Remember that we will probably meet again; we should remember the ongoing relationship, and should act with trust in the negotiation.

The Stages in the Process

Most of us are problem-solving people; we are all eager to solve problems, and to do that as quickly as possible. We look for the 'one minute' solutions.

I suggest that the one-minute negotiator has about as much success and has a life that is about as satisfying as the one-minute lover. Negotiation is a slow process, and we need to hold back from our eagerness to get to a solution. That goes against much of what we get rewarded for in management, where we are encouraged to seek quick solutions. Negotiation, on the other hand, requires patience and an open mind. It involves learning, getting the facts right, articulating our interests, articulating the other side's interests, and often backing away from our 'pet' solutions or outcomes.

One metaphor in negotiation is a win/lose one, borrowed perhaps, from sporting contests: 'we'll go in there and thrash them'. Or it could be a win/win metaphor: 'let's see what we can work out together; I have nothing firm in mind'.

The process I suggest has five sequential stages:

1 mating dance
2 diagnosis
3 exploration of interests
4 problem solving
5 commitment to outcome

Trust and Ongoing Relationships

This is, perhaps, the most difficult concept in the process. Most of us think of trust as a moral precept. But we can be more hard-headed about trust. Thinking about the ongoing relationship:

- What will be the consequences for them if they renege on the deal?
- They pull a swiftie now, may I get the chance to do likewise in the future?
- Do we have some way to make contracts legal and binding?

In the introductory session, we take part in a 'tragedy of the commons' simulation. This illustrates the importance of trust in decision making, and how our strategy in planning must address the issue of trust. From a shared commons, everyone can draw what they need, so long as each person trusts all the others not to exceed their allotted share. As soon as anyone does so, there is no longer enough for anybody.

Secrecy in Negotiations

Negotiations proceed well if the core stages (diagnosis, exploring interests, problem solving) take place behind closed doors, with no record. That makes it easier for parties to confront the problem, not the people. It becomes easier to suggest options without commitment. It is for this reason that there is a paucity of good negotiation case studies. What goes on behind the doors is best if it is as informal (but well disciplined) as possible.

Adversarial and inquisitorial processes are ways of resolving disputes, but they are not negotiated processes. They require an adjudicator and, in nations with a modicum of commitment to democratic process, are conducted in the open. Sometimes, in civil and industrial relations proceedings, the dispute resolution may sally back and forth between negotiated and judicial processes, but the processes are different.

Sometimes negotiating parties may appoint a mediator. This person is not an adjudicator, but is a facilitator, helping the parties go through a negotiation process, and often, as is the case with marriage guidance counselling, helping separate the people from the problem.

Conflict Resolution Styles

This exercise has been adapted from Watson et al, *Structured experiences and group development*. The proverbs listed below reflect traditional wisdom for resolving conflicts and raise a number of different confllict-resolution approaches.

Read each of the proverbs carefully. Using the scale, indicate in the brackets how typical each proverb is of your actions in a conflict.

5 – Very typical of the way I act in a conflict
4 – Frequently typical of the way that I act in a conflict
3 – Sometimes typical of the way I act in a conflict
2 – Seldom typical of the way I act in a conflict
1 – Never typical of the way that I act in a conflict

		Score
1	Use soft words to win hard hearts	[ ]
2	Come now and let us reason together	[ ]
3	For your arguments to have weight, argue loudly and forcefully	[ ]
4	You scratch my back, I'll scratch yours	[ ]
5	The best way to handle conflicts is to avoid them	[ ]
6	When one hits you with a stone, hit them with a piece of cotton	[ ]
7	A question must be decided by knowledge and not numbers if it is to be a right decision	[ ]
8	If you cannot make a person think as you do, make them do as you think	[ ]
9	Better half a loaf of bread than no bread at all	[ ]
10	If someone is ready to quarrel, they aren't worth knowing	[ ]
11	Smooth words make smooth ways	[ ]
12	By digging and digging, the truth is discovered	[ ]
13	He who fights and runs away lives to fight another day (once you strike, withdraw so that you can strike again)	[ ]
14	A fair exchange brings no quarrel	[ ]
15	There is nothing so important that you have to fight for it	[ ]
16	Kill your enemies with kindness	[ ]
17	Seek until you find, and you'll never lose your labour	[ ]
18	Might overcomes right	[ ]
19	Tit for tat is fair play	[ ]
20	Avoid quarrelsome people – they only make life miserable	[ ]

CONFLICT RESOLUTION SCORE SHEET

Conflict strategy

	Proverb no	*Score*
Avoiding	5	_____
	10	_____
	15	_____
	20	_____
		Total _____
Forcing	3	_____
	8	_____
	13	_____
	18	_____
		Total _____
Smoothing	1	_____
	6	_____
	11	_____
	16	_____
		Total _____
Compromising	4	_____
	9	_____
	14	_____
	19	_____
		Total _____
Problem solving	2	_____
	7	_____
	12	_____
	17	_____
		Total _____

Note: The higher the total score for each conflict strategy the more frequently you tend to use this strategy. The lower the total score for each conflict strategy, the less frequently you tend to use this strategy.

Discussion

Every strategy is appropriate under some circumstances:

- **Avoiding** is unassertive and uncooperative – the individual does not immediately pursue his or her own concerns or those of the other person. He/she does not address the conflict. Avoiding might take the form of diplomatically sidestepping an issue, postponing the issue until a better time, or withdrawing from a threatening situation.
- **Forcing** is assertive and uncooperative – an individual pursues their own concerns at the other person's expense. This is a power-oriented mode, in which one uses whatever power seems appropriate to win one's own opposition – their ability to argue, their rank, economic sanctions. Forcing might mean standing up for your rights, defending a position which you believe is correct, or simply trying to win.
- **Smoothing** is unassertive and cooperative – the opposite of forcing. When smoothing, an individual neglects their own concerns to satisfy the concerns of the other person; there is an element of self-sacrifice in this mode. Smoothing might take the form of selfless generosity or charity, obeying another person's order when they would prefer not to, or yielding to another's point of view.
- **Compromising** is intermediate in both assertiveness and cooperativeness. The objective is to find some expedient, mutually acceptable solution that partially satisfies both parties. It falls on a middle ground between forcing and smoothing. Likewise, it addresses an issue more directly than avoiding, but does not explore it in the same depth as problem solving. Compromise might mean splitting the difference, exchanging concessions, or seeking a quick middle ground position.
- **Problem solving** is assertive and cooperative – the opposite of avoiding. Problem solving involves an attempt to work with the other person to find some solution which fully satisfies the concerns of both persons. It means digging into an issue to identify the underlying concerns of the two individuals and to find an alternative which meets both sets of concerns. Problem solving between two people might take the form of exploring a disagreement to learn from each other's insights, agreeing to resolve some condition which would otherwise have them competing for resources, or confronting and trying to find a creative solution to an interpersonal problem.

We might illustrate the danger of fixed conflict management styles by the fable about the two sisters who were fighting over an orange.

- In avoiding the conflict, both sisters leave the orange rotting in the fruit bowl. For one to claim it would raise a conflict with the other.
- In smoothing, each sister says to the other 'You take it', to which the response is 'No, I couldn't'. The orange still sits in the bowl.
- Forcing, simply means that the stronger sister takes it – wrenching it from the hand of the other, or possibly using some threat.
- A purely compromise solution would be for each sister to take half the orange.
- A problem-solving solution would be for each to discuss with the other why she wants the orange. So goes the fable that one sister wanted the skin to make a cake, the other the inside to make juice!

Tragedy of the Commons: the Developers of ID

Exercise designed by Ian McAuley, Consultant, Schlich Street, Yarralumla, Canberra, Australia

Rules

In groups of seven to nine people, each person represents a different city in the same region. One member acts as the developer whose aim is to bring great profits and advantages to the region, and leads the exercise.

Background

The region of Id has five cities of around 40 000 each, strung along either side of a river. Id is an old industrial region. Once host to a number of 'smoke-stack' industries, it has seen rapid de-industrialisation over recent years, with a consequent rise in unemployment – now running at 15 per cent.

The environment is not in good shape, but none of the local city councils have the funds to devote to urgent environmental projects – reclamation of disused mine sites, replanting trees, clean-up of abandoned factories, moving old dumps, etc. There is, however, the chance to get investment projects in the area. The region could easily take up to two projects, with net employment and environmental benefits to the region. Although each project would add pollution, that addition would be incremental, and would help the city concerned raise taxes to pay for urgent environmental work. The benefits would accrue not only to the host city, but also to the region. However, if more than two projects in the region were established, the costs would start to outweigh the benefits.

Your Role

You are a small executive government group for your city, and you must decide whether or not to accept a project in the coming period. The benefits (or costs) depend not only on your decision, but also on what the other cities do. Environmental scientists, engineers and economists have come up with a schedule of benefits for your city associated with accepting a project as shown below. These benefits are in ordinal terms; you may wish to think of them as millions of dollars, but they add, as far as possible, economic and environmental benefits (and costs) – see Table 10.1.

Table 10.1   Schedule of benefits for my city

Decisions of other cities	Decision of my city	
	Accept a project	Reject a project
If no other cities accept	100	0
If one other city accepts	60	-10
If two other cities accept	0	-30
If three other cities accept	-20	-50
If all others accept	-40	-80

The Simulation

You are at a meeting of representatives of each city where the developers will be offering projects to the cities.

You will have a number of rounds, of a few minutes each. In each round the prospective developer will ask for bids, and will take all bids offered. The developer will count down, and at the end of the countdown, you will show a card – DEVELOP or DON'T DEVELOP. Your objective, as an elected official, is to do as well for your city as possible. You may not collude with other cities. You can keep your score below:

Round	Accept/ Reject	Others accepting	Score	Cumulative
1				
2				
3				
4				
5				
6				
7				
8				
9				
10				

Discussion

As in each of the simulation exercises, the discussion should take about as long as (or longer) than the exercise. After the end of the simulation, participants discuss how and why they bid as they did. The lessons learnt should be listed. What would have happened if, halfway through the bidding, all of the participants had been allowed to collaborate with one another, and draw up a bidding strategy?

'Tragedy of the commons' situations abound. They form a class of situations where:

- If you and I are both selfish we will look after ourselves, but will not be as well off as we would if we cooperated;
- If you and I cooperate we shall both be better off in the long run, and we will also maximise our resources for the future;
- If you act as if I am going to cooperate, but I act selfishly, I will do very well, and you will come off very badly.

We can think of situations such as a refrigerator in a student dormitory. If we all simply look after ourselves, and steal a cold bottle of beer whenever we find it, all the students will keep their beer in their rooms, warm. The refrigerator will stay unused, unmaintained and will eventually be vandalised. If most people are prone to stealing cold beer, but some suckers put their beer in the refrigerator, the thieves will get cold beer, and the suckers will lose. They will soon learn not to keep their beer in the refrigerator. The thieves can use the refrigerator all they like, but they will get no more free beer. If the whole group agrees not to steal beer, then the situation may stabilize, with a social contract existing among the group. But it could become unstable. Unless there is guidance from higher authority – eg a moral code or sanctions – the situation could unravel over a period of time.

Another application is in international trade theory. If all countries maintain trade restrictions, then they are not as well off as they would be with more open trade. However, if just some countries liberalise, then they run the risk of exploitation.

The value of social contract without sanctions is dependent on the existence of ongoing relationships. Further, a social contract is more easily applied in a small group than in a larger one. It is important to note, however, that there is no automatic mechanism whereby a social contract arises. In the absence of outside authority it requires a conscious act of negotiation within the group.*

It would be contradictory to give a set rule for behaving in 'prisoner's dilemma' or 'managing the commons' situations. However, research has shown that those who maximise their scores over many rounds tend to use a strategy along the following lines:

1   Be cooperative to start.
2   Be frank and responsive to negotiation.
3   Be forgiving – give everyone a second chance.
4   Be clear about the rules.

Brainstorming

Brainstorming, or idea generation, can (and should) be a structured activity, carried out early in the life of a project, in response to a significant problem, or in difficult negotiation situations when both sides are aware of each others' principles, but no solution is evident. It is usually enjoyable, but it is not a free-for-all. There are rules which will help get the most out of the experience.

To allow participants to concentrate on the task of idea generation it may help to appoint a non-involved facilitator. Besides enforcing the 'no criticism' rule, the facilitator can also encourage participation of quieter members.

Rules for Brainstorming

- No evaluation of any kind:
  - Put energy into generating ideas, not into defending them.

---

* Those associated with game theory will recognise this as a multi-party 'prisoners' dilemma' simulation. It is also a simulation of Hardin's 'tragedy of the commons'. Hardin, G (1959) 'The tragedy of the commons' *Science* **162**, pp 1243–8

- Think of the wildest ideas possible:
  - Suppress internal judgement which may inhibit new ideas.
- Go for quantity of ideas:
  - Quality can come later.
- Build on ideas of others:
  - Combinations or modifications of previously suggested ideas often lead to new ideas superior to the original ones.
  - Wacky ideas may help more sober people break out of limiting.
- Ignore rank, qualifications and other indicators of authority:
  - Good ideas can come from unexpected quarters.
- Keep a record:
  - It's easy to forget.
  - Keep it open, and don't put names against ideas.

Brains form the potential applications of each exercise to environmental management of local issues throughout the course.

11 Stage 1: Exploration and Forecasting

All of the issues on the table

LEARNING MATERIALS:

- Forecasting long-term trends
- Guided imagery
- Futures wheel

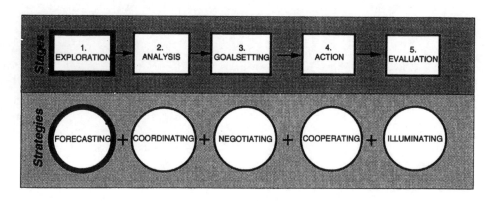

Figure 11.1   All of the issues on the table

AIMS AND OBJECTIVES

- To trial three different methods for exploring the future.
- To allow for new ideas and innovative perspectives.
- To separate the issues from the people involved.
- To identify all the sources of information on an issue, whether explicit, implicit, or missing.

In the first stage of environmental decision-making, the broader and more complete the coverage of the issues, the better for future management. The task in this session is to explore future trends for a selected issue. Chapters 6–9 contain case studies of conservation, costing and rehabilitation of environmental resources. In each case, all dimensions of environmental problems will need to be considered – including legal, political, economical, administrative, scientific and community interests – in order to provide all the elements needed in the problem solving.

The challenge for everyone in a prospective management team is to arrive at a shared, cooperative perspective on the issues involved. The exercises in this session are designed to help members of the group to achieve this perspective. The ground rule for this session is to accept all ideas and options at face value, as they are offered, without criticism or rejection from any of the team members. Only questions of clarification should be permitted.

There is always a great deal more information available on an environmental issue than any one person can handle. The information may be **explicit,** in which case it is obtainable from files and records. Information used in management is often **implicit**; that is, it is not recorded formally, but is well known to those who work together. There is normally a pool of **missing** information, that no-one has yet recognised will be needed.

In environmental management, each expert group is often unaware of the information held by others. It is all the more important to ensure that all the issues are on the table from every perspective, before assuming that any type of information is unavailable or irrelevant.

The three exercises in Stage One of environmental management allow for the development of;

1   a broad view of the future local, national and international pressures for one environmental issue;
2   against this background, a particular issue is then explored in depth by visualising the issue in the future; and
3   the implications for environmental management of the issue are identified by the group.

RISKS AND OPPORTUNITIES

In exploring all the possibilities for environmental decision-making, the ideal is to make the forecasts of future action working as a team in an open-minded, constructive atmosphere. This provides the maximum opportunity for generating new and constructive ideas. This is harder than it may sound in critical, competitive societies. Risks to be avoided at all costs are allowing existing prejudices to be admitted (otherwise the whole process may be set in concrete), or letting the brainstorm float off so far from reality that suggestions might as well come from outer space.

The necessary conditions for success in this type of brainstorming session are:

- A facilitator who:
  - makes sure everyone is involved;
  - stops anyone hogging the limelight;
  - does not allow criticism;
  - helps people share their ideas;
  - keeps a running record of the ideas (but not the people).
- Ground-rules which:
  - make it safe for anyone to suggest anything;
  - put all the energy into the ideas not the reasons behind them;
  - encourage everyone to build on each other's ideas;
  - allow wild ideas to be developed.

Three issues have been documented for the purpose of providing realistic course materials for each stage of environmental decision-making. Each has its own case study; each represents a different management issue within a different mode of political economy. The challenges in each of the hypothetical case-studies are:

- Rehabilitation of environmental degradation due to overuse of subsidised water resources (Saxa Shire) – Stage 1.
- Conservation of water resources against uneven sources of supply using cooperative management (Trickle Creek) – Stage 2.
- Open forum on integrated response to traffic gridlock (Bangkok) – Stage 3.

ACTIVITIES

Theme for Forecasting: Water as a Public Resource

Each activity can be undertaken at weekly intervals, or a single one can be selected for a short course.

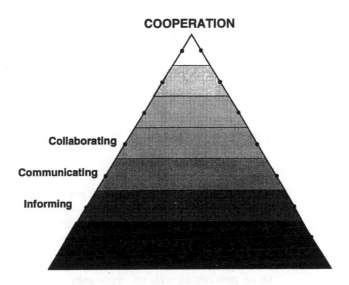

COOPERATION

Collaborating

Communicating

Informing

Figure 11.2 All the issues on the table: activities

- **Long-term trends** (150 mins). This exercise places a selected issue (such as water resources) within the context of global issues and local social values and expectations.
- **Visualising the future** – guided imagery (120 mins). The use of creative imagination in future searches is becoming more and more common as it becomes clearer that current institutions alone are not providing the solutions for stable environmental management. This activity requires concentration and projection of ideas beyond current limitations and constraints.
- **Exploring the issue** – futures wheel (60–120 mins). Issues already identified through activities such as brainstorming or future searches are developed toward comprehensive practical strategies for action.

The three exercises are aimed at, in turn:

- **Informing**: putting the issues in a social and global context and within a time perspective;
- **Communicating**: sharing information on all the element of an issue;
- **Collaborating**: helping the group develop practicable shared goals. For details of each exercise, see the Course Coordinator's Guide.

199

This learning module applies the first of the four stages in conflict resolution described in Fisher and Ury's *Getting to Yes* – the separation of the people from the issues.

MATERIALS

Forecasting Long-term Trends: Helping a Team to Plan the Future

Environmental Futures

Ecologically sustainable development is a challenge to management for both the present and the future. Issues management is a means to assess the future and relate the findings to a specific issues programme.

There are three good reasons to study the future. First, it is a sensible precaution. Second, it provides a base for personal, corporate, collective and government planning. Third, and most significantly, an effective look into the future provides us with an opportunity to control, to manage, and in the environmental area, to ensure the future.

Most, if not all, serious futurists share four fundamental prepositions about the future:

1   There are real alternative futures ahead of us.
2   We can, to a useful extent, anticipate and describe what those futures may be.
3   We have the capacity to chart a course amongst those alternative futures rather than just drift in a tide of time and events.
4   We have a professional and personal obligation to use these capabilities.

Our World Has Changed

Another reason for the study of the future is that the world has changed in the past two decades, to an extent that makes obsolete many of the traditions, views, orientations, theories and policies about the collective and social management of our public and private worlds.

Sixty long-term trends are outlined at the end of this section. Together they constrain, define, and form the basis for our future. One can epitomise those changes in some brief statements. The most fundamental of these statements is that people in all the industrialised countries today live in an almost totally human-made world. For most of us, little that we encounter, whether in eating, sleeping, relaxing, or working, is *not* the product of human enterprise. What is important about that is that we know far less about the stability, strength, resilience, healing power, or adaptability of that human-made world, than we do about those aspects of ecological systems. We know far less about the technological impact of the human-made world compared to what farmers in a non-industrialised country know about their natural world. Fundamentally, we do not understand enough about this world we have created in the last few decades and which we are continuing to rapidly expand and elaborate.

Technology, in every society of the world, is now all-pervasive in its scale, scope, interrelatedness, pervasiveness, the size of its investment, rapid turnover, and its universal impact on every aspect of lives. This new role for technology has led us to what Daniel Bell calls the post-industrial society. This is not just jargon, but a fundamental conceptual theme that has to do with a basic shift in the structure of our society and economy. At the beginning of the last century, the bulk of the work force was in agriculture. Systems of environmental management included the feudal (ownership of land and labour) and the commons (social customs and rules ensuring stability of use). By the beginning of this century, the industrial revolution was fully in place, and with it the management of industrialised societies through monetary exchange and economic criteria. About 15 years ago the chief activity of the labour force made a transition predominantly into services. For example, in Australia, perhaps as little as 45 per cent and as much as 65 per cent of the labour force, now deal with the new dominant commodity of society. This is not agricultural products or petroleum. Neither is it automobiles, housing, or transportation. The dominant commodity of modern post-industrial society is information. Science, government, media as sources of knowledge, are now the central drives in the organisation of society.

A fundamental, ironic, and intrinsic feature of the new knowledge is that it automatically generates new ignorance. The more striking the new knowledge, the more fundamental its ramifications, the wider and deeper the pool of ignorance which it generates. Consider development in telecommunications, energy use and genetics in the past three decades. The systematic exploration of the future is more and more urgent. As our public competence and command of information expands, the more urgent becomes the common pooling and sharing of the new knowledge.

Forecasting a Sustainable Future (180 mins)

Groups of five to nine people brainstorm the implications for environmental management for each of the following sixty aspects of future change. The brainstorming should be fast and furious, with one of the group acting as reporter, recording suggestions for all to share.

The process should be about 15 minutes for each of the five sections: trends in society, technology, workforce, values, family and institutions. For a further 15 minutes each group should review the aspects of environmental impact of the changes they have recorded, and identify the most significant for the future in each section.

After a break of 15 minutes, groups come back together to compare notes. Free discussion of the implications for environmental management is summarised as key points by a recorder (30 mins). When a priority list has been established by the full group this then acts as a guide to the issues to be examined in the next stages of the environmental management process.

General long term societal trends

1   Economic growth of capital, and inflation
2   Expanding educational levels throughout society
3   Rise of knowledge industries and a knowledge-dependent society
4   Relative decline in common knowledge of the physical world
5   Urbanisation/metropolitanisation/suburbanisation
6   Increasing gap between rich and poor, within and between countries
7   Cultural homogenisation – the growth of a national society
8   Growth of a permanent military establishment
9   Mobility – personal, physical, occupational
10   International affairs and environmental security as a major societal factor

Technology trends

11   The centrality and increasing dominance of technology in the economy and society
12   Integration of the national economy
13   Integration of the national with the international economy
14   The growth of research and development as a factor in the economy
15   High technological turnover rate
16   The development of mass media in telecommunications and printing
17   An awareness of the finite nature of resources
18   An awareness of the finite nature of waste disposal

Trends in labour force and work

19   Specialisation
20   Growth of the service sector
21   Relative decline of primary and secondary employment
22   Growth of information industries, movement toward an information society
23   Rise in managerialism
24   Expansion of credentialism
25   Women and disabled entering the labour force
26   Early retirement

27 Unionism
28 Growth of pensions and pension funds
29 Movement toward second careers and midlife change in career
30 Decline in the work ethic
31 Rising unemployment, unequally distributed throughout society

Trends in values and concern

32 Diversity as a growing, explicit value
33 Decline in authority as an authority
34 Consensus and collaboration as explicit values
35 Increasing aspirations and expectations of the rewards of success
36 Growth in tourism, vacationing and travel
37 Expectations of high levels of medical care and social services
38 The growth of consumerism
39 Growth of physical culture and personal health movements
40 Civil rights, civil liberties expansion for blacks, gays and other minorities
41 Growth of women's rights and responsibilities

Family trends

42 Decline in birth rates
43 Shifts in rates of family formation, marriage, divorce, and living styles
44 The growth in leisure
45 The growth in the self-sufficiency movement
46 Improved nutrition with the consequent decline in the age of puberty
47 Protracted adolescence
48 Decline in the number and significance of rites of passage, birth, death, marriage etc
49 Isolation of children from the world of adult concern
50 Age-restricted peer groups
51 The growth of a large aged population
52 The replacement of the extended family by the nuclear family, living singly, and other living arrangements

Institutional trends

53 The institutionalisation of problems. This is the tendency to spawn new institutions and new institutional mechanisms for dealing with what were in the past personal, private, or nongovernmental responsibilities
54 Bureaucratisation of public and private institutions
55 Growth of big government
56 Growth of big business
57 Growth of multinational corporations
58 Growth in future studies and forecasting and the institutionalisation of foresight mechanisms and long-range planning
59 Growth of public participation in public institution and private institution decision making
60 The growing demands for accountability and expenditure of public resources
61 Growth of demands for social responsibility

Guided Imagery: A Vision of a Sustainable Environment

Aim

This exercise allows a group to develop their goals for a shared future, and prepare a coordinated strategic plan or action agenda. The exercise has been widely used in the World Health Organisation Healthy Cities Project and was designed by Trevor Hancock, Public Health Consultant, Klemberg, Ontario.

For the purposes of this exercise, the environment can be either a real place the group will work in, or based on the description of an area constructed during the Futures Wheel exercise. The following example has been written for the case of Nyah Shire, although any of the environmental case management studies may be used.

Spend 2–3 minutes discussion on: *What would people most like to experience in a healthy, sustainable environment?*

Make sure that the group is in a quiet room, with no interruptions. In a clear, slow, calm, deep voice read the following script. Give time for people to develop the images in their minds. Some people will have greater difficulty visualising than others. This does not really matter; they can still contribute to the activity later. Pace the exercise yourself. Leave plenty of time for everyone to visualise the images, and move slowly from one set of images to the next.

'We are now going to take a trip to the future, to an ideal sustainable Saxa Shire (or Trickle Creek or Smoke City) about twenty years from now. This is not Saxa Shire as it is today, nor as we think it probably will be – this is Saxa Shire as we would like it to be if all our expectations had been fulfilled and we had achieved a healthy sustainable environment. All the problems of today have been solved.

Make yourself comfortable – you may find it useful to close your eyes so that you can more easily imagine the landscape (or region, or city) in your mind's eye.

I want you to imagine that you are hovering high above the principal town in Saxa Shire; perhaps in a balloon or helicopter – an ideal Saxa about twenty years hence. Look down at the landscape beneath you. What does it look like? What colours and shapes do you see? What time of year is it? What would it look like at a different time of year? Look out across the landscape at the edges of the town, the paddocks and farms, the hills and valleys. What crops are growing? What stock is feeding? Where does their water come from? How is the water distributed? How pure is the water? What is it used for? Where does the waste water go? Where are the farms? Can you see people and goods moving to and fro? How are they being moved? What sounds come up to you from farm and town? What smells?

Now I want you to descend. Look around you as you come down into the middle of Saxa's main town. Look at the shapes and structures of the buildings. Look at the spaces between them. As you get closer to the ground, listen to the sounds and smell the scents of the landscape.

Are there people around? Who is there? What ages are they? How do they react to you – and to each other? What activities are going on? Are there people working? Are there people relaxing? What else are they doing? Who is doing it? Now walk around and see what else there is.

Join one of the people as they move about. What are they doing? How do they look? Is this a good place to be? How are the people interacting with one another? Now join some other people. What is happening here and how does it feel?

Now imagine yourself on a farm. How is this different to the situation today? What is happening? What is happening in the neighbourhood and in the community? What are people happy about? What are they worried about?

Imagine that it is lunchtime. Where do people go for lunch? What do they talk about? What sort of food are they eating? Where did it come from? What do they do after lunch?

As the day comes to an end, go home with someone. How do they get home? Where is home? How far are they travelling? Go with them in their neighbourhood. Remember, this is an ideal, well-managed sustainable environment. How does the neighbourhood feel? Do you feel safe? Do you feel secure about the future? How is the region being managed? Where are the decisions being made? What important changes have happened recently?

Now go into the house. Who is there? Is it a family? Who is in the family? Are they all earning their living? What do they do? What about the others? Are there children? Do they go to school? What are they learning about their district and their environment? Eat dinner with the household. What do they talk about? What do they do for relaxation?

Now imagine it is a weekend. Pick any season of the year you like. Move around the district. What are people doing? What recreation do they have? Where are they going? Is anyone working? What are they doing? Now imagine it is a different season. What is happening in this season? Is the work different? Is the recreation different?

Now, before we leave this ideally well-managed district, think back on all you have seen. Think about how the country looked, who was in it and what they were doing. Did you see the very young and the very old? Where were they? How was life for them? Did you call on the Shire Office? What were they doing? What were their plans? Do the issues that we discussed earlier appear to have been addressed? What still needs to be done? Who is meeting the challenge?

Now, I want you to come back to the present time, reflecting on all you have seen. (*Provide time here to return to the here and now*). When you are ready, write down the ten most noticeable things you saw, heard, smelt or touched. Write down things that surprised you, the things that pleased you, the things that worried you.'

Give the participants five minutes or so to write down the ten things that stood out for them. If necessary put some pressure on the participants to go to ten items – the last two or three may be difficult, but often contain the most original ideas.

Spend another five minutes discussing the ten issues that people have written down. What did people see and hear? Did they find it hard to do? Don't let any of the participants be critical of anyone else's issues – although they are allowed to ask to have the topic clarified.

There will always be someone who is very practical, or not at all visual, who has trouble doing the exercise – that's fine. They should be made to feel comfortable, and asked to contribute to the next stage of the exercise from guesswork or their own experience, rather than the guided imagery.

After this short discussion, have each person write on three cards, in five words or less and in big letters that people can read across the room, the three most important ideas among among their ten. Each person is asked to stick their most important card of the three up on the wall themselves, and then read out the idea. Others may ask questions of clarification, but they may not offer comments or criticism. Each person should either put their card in a group with other issues, or start a new category.

Then take a second round, then the third. No discussion and no criticism of any contribution is allowed. Short questions of clarification are permitted. Everyone puts their cards wherever they wish, and no-one can move another person's card without permission.

When the cards have been put into a minimum number of groups (probably three, perhaps five) each strand should be given a name/title encapsulating the central idea. This then gives the central conditions for sustainable development in Saxa Shire.

Futures Wheel

About This Activity

By creating futures wheels, we can diagnose how developments in one area will automatically lead to developments in other areas. We can begin to visualise the impact of a forecast in one area on another. Only if all the interacting are included in the problem-solving will the solutions be long lasting.

What You Need

Large sheets of papers and coloured pens

How to Do It

First agree as a management group on the issue or problem that is to be faced, eg acute rise in water costs, pollution of water supplies or oil dependency as in Figure 11.3.

Brainstorm the impact this may have on other areas. For example, 'less pollution', 'alternative energy', 'unemployment' and so on. Do a sample futures wheel to test out the scope of the issue.

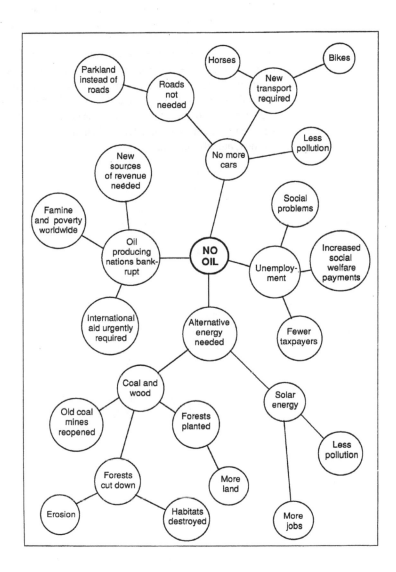

Figure 11.3 Futures wheel

1 Decide on a forecast, selected from the brainstormed list in the centre circle. Treat the forecasting as a creative, open-ended exercise, not a prediction of certainty.

2 Write down forecasts which will result directly from the initial forecast. Join the second ring of forecasts to the first with 'spokes'. Continue the forecasts out to a third fourth, fifth ... ring of forecasts.

3 When the system seems complete, identify which forecasts are related to each other and circle them in the same colour to identify them. This will demonstrate that the effects of change can be both direct and indirect, but still related.

12 Stage 2: Analysis and Coordination

All the players around the negotiating table

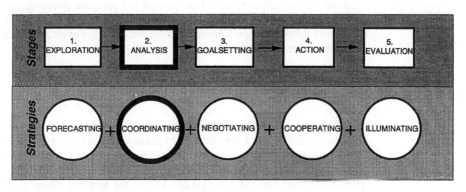

Figure 12.1 All the players around the negotiating table

AIMS AND OBJECTIVES

- To identify key players and resources in an environmental issue.
- To coordinate key players and resources so that they work constructively, not competitively.
- To identify the interests and bargaining styles of the various stakeholders.
- To form a fully representative management team.
- To simulate a strategic planning meeting.

This session covers the second stage of management decision-making after the main players have decided on the full scope of the issues. The resulting complexity has to be pulled into some sort of manageable order. The skill in identifying the key interests, players and resources will determine the success of the management – whatever position an individual holds within the management team.

The management team needs three different sorts of people: technical experts relevant to the issue, representatives of the stakeholders' interests, and people to coordinate the management team. The usual method of bringing these together in Stage 2 of environmental management is to form a management committee.

In practice, the committee may be called a working group, a panel, a council or an advisory group. Whatever it is called, the same rules apply. Someone – the chair, coordinator, convenor or facilitator – is needed to ensure fair play. This is more important in environmental decision-making than in almost any other area. The environment has more separate interest groups than any other field of social policy. Chairing a committee is one of the most common – and most taxing – forms of conflict management.

Notes below provide some light-hearted but practical guidance on how chairing groups can be a constructive enterprise. The task in this session is to coordinate the players, the interests and the resources in designing a mutually productive management strategy.

RISKS AND OPPORTUNITIES

In this session, the task is to coordinate the key interests in a way that reduces the risks of the lunatic fringe or the old guard taking over – or worse, of the whole committee being locked in a stand-off between the two. A working group is formed to plan, manage and evaluate action on the issues identified in the previous session. Selecting the membership of the working group requires a careful analysis of the stakeholders and resources available, followed by agreement between members on the rules for working together.

In choosing the following, key interests need to be represented in almost any environmental decision-making:

- major political interests;
- major administrative interests;
- research interests relevant to the issue;
- local and national financial resources;
- community and consumer interests;
- individuals who are key players; and
- key services and agencies (eg engineers, planners, lawyers).

In deciding how to run the management committee, the following are good practical rules:

- decide on the ground rules together at the first meeting;
- write them down and agree to them;
- allocate tasks of chair (facilitator) and secretary (administrator);
- set a mutually agreed agenda, remembering that the person who controls the agenda controls the debate;
- become familiar with committee rules and processes (eg *Preparing for the Meeting* in the Course Coordinator's Guide, Appendix 1); and
- read *Getting to Yes* (especially *Negotiating on Merits*, p 13).

ACTIVITIES

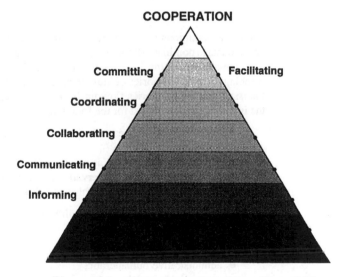

Figure 12.2   All the players around the table: activities

Theme: An Integrated Environmental Management Plan

1 Stakeholder analysis
2 Power and resource analysis
3 Committee procedures: preparing for the meeting

The exercises address the skills of:

- **Collaborating**: identifying key players and their resources;
- **Coordinating**: setting up a working group;
- **Commitment**: agreeing on a joint strategy;
- **Facilitating**: chairing a working group or committee;
- **Communicating**: sharing the learning between the key players.

These activities allow participants to practise the second stage in the Harvard Negotiating Process – identifying interests, not positions.

MATERIALS

Stakeholder Analysis: Including All the Interests

Background

Set up the conditions for productive brainstorming as in Stage One, invite the participants to nominate all the stakeholders in a selected issue. It may be useful to nominate one of the issues selected above. There will be no problem in reaching 30 or 40 categories of individuals and organisations with an interest in the issue. Let the categories flow unchecked. After the group has constructed the initial list, check the list for the following:

- consumers of services and clients of agencies as well as providers and managers;
- private interests as well as public bodies;
- industries as well as conservation groups; and
- future needs as well as established interests.

People from the following categories will be needed as members of a management group:

- technical/scientific;
- administrative/bureaucratic;
- economic/financial;

- policy/political;
- regulatory/legal; and
- community interests.

Since an effective management group should be not less than five (or else there would be too few to carry the responsibilities) or more than twelve (too many to coordinate efficiently) the next step is to select between five and twelve key stakeholders who will between them satisfy the requirements listed above.

Assume for the purposes of the rest of the exercises that the management group is between seven to nine. This exercise is completed when the group has agreed on a seven-member management group (or however many members there are in the learning group) which fits the criteria listed above.

It will be important to choose people who wear more than one hat, and bring with them the maximum resources. The exercise which follows is an analysis of the power and resources of the key stakeholders.

Power and Resource Analysis (30 mins)

Participants from groups of seven to nine people. Each group takes a priority issue identified in the proceeding session on exploration, first analyses the components and then integrates the interests into a common direction.

Analysis

Everyone in a community has some power and resources to contribute to resolving an environmental issue even if this is not immediately apparent. Consumers vote, pull strings, and evaluate their environmental quality as much as the official environmental managers.

There are four bases of power in all societies, from the most authoritarian to the most equitable. In every environmental issue the participants will have access to resources based on:

- **who you know** – socio-political power (a formal position, informal networks, social standing, personal and work relationships, community service political positions and loyalties);
- **what you own** – economic/financial power (money, ownership of goods or services, anything valued by the market economy, even debts);
- **what you know** – technical or theoretical knowledge and skills (formal and informal sources of information and skills; theoretical and practical understanding of an issue, control over knowledge of others as in the education system) and;

213

- **who you are** – each individual's personal experience (level of skill, insights, previous performance, work roles, private roles, personal reputation, professional roles).

Activity: cross-reference each of the key stakeholders and each of the four bases for power. All stakeholders will have some measure of power in all categories. The effect of this information shared within a group is to demystify power; to empower members who felt powerless; and to allow the group to start on a fairly level playing field. Record results on table 12.1.

Discuss the insights into the potential for the key stakeholders to work together. Do alliances and checks and balances to equal power become apparent?

Table 12.1   Stakeholder resource analysis grid

Stakeholders' power bases

	Social	Economic	Technical	Personal
Stakeholder				
1				
2				
3				
4				
5				
6				
7				

Discussion of Interests

As you join a committee meeting, out of all the interests in the main issue, try to understand what the parties really want.

Working assumption: focusing on interests rather than positions increases our chances of achieving a good outcome.

1   We tend to focus on positions, not interests. If we assume that a negotiation problem consists of a conflict of positions, then it makes sense to talk about positions. This is a common tendency. But the basic problem in negotiation lies not in conflicting positions, but in the conflict between each side's needs, desires, concerns, and fears. Such desires and concerns are *interests*.

2   We assume that all of our interests are opposed. When we focus on positions, we can assume that because the other side's positions are

opposed to ours, their interests must also be opposed. If we have an interest in defending ourselves, then they must have an interest in attacking us. In many negotiations, however, a close examination of the underlying interests will reveal the existence of many more interests that are shared or compatible than one that are opposed.

3   Focusing on interests facilitates creative problem-solving. For every interest there usually exist several possible positions that could satisfy it. When we look behind positions for the motivating interests, we can often find an alternative position that will satisfy our interests as well as theirs. For example, a nation may assert the position that a border must be drawn at a certain place; interests which lie behind this position may include national security, access to mineral resources or questions of sovereignty. If the parties look behind their positions and focus on their interests, an agreement may be reached which reconciles seemingly contradictory positions. One nation could retain sovereignty over the land, while the other could retain the rights to the mineral resources. It is far easier to accommodate interests into a mutually acceptable package than it is to accommodate positions.

4   Consider the following guidelines:
   - *When preparing, focus on interests.* First, clarify our own interests. Second, and perhaps most important, try to understand the interests of the other side. This will increase our understanding of the problem and will help us invent solutions which meet not only our interests, but the interests of the other side as well. One way to uncover their interests is to examine from their point of view what it is you want them to agree to (their 'currently perceived choice'), and then determine what interests of theirs are preventing them from being able to agree to it.
   - *Focus the negotiation discussion on interests, not positions.* It is difficult to fashion a creative solution to a problem which satisfies the interests of different parties if the interests of each are not explicitly discussed.
   - *Think of positions as clues.* If they continue to talk about positions despite your efforts to the contrary, ask them for help in understanding what is leading them to this position. Ask them 'Why?'
   - *Use leadership.* Be prepared to take the lead by talking about some of your own interests. If you are not willing to tell them something about your own needs, desires, concerns, and fears, then why should you expect them to be willing to do the same?

Developing a Policy (60–180 mins)

Definition of policy (Titmuss): the principles which direct action towards predetermined ends.

A committee or a working group is the most usual method of developing environmental policies, strategies or action plans. This is more likely in the environment area than any other because of the complexity of both the expert advice needed and the machinery to implement the action.

Committees are often described as a waste of time; yet they are designed, and have functioned for a centuries as a primary mode of conflict management. Skilled managers can use a committee to achieve almost any desired result – including no result. Practical advice on committee procedures is included in the readings for this section.

Set up a committee made up of the seven central players selected from the stakeholders in the environmental issue. Everyone now knows the resources each commands.

The group should proceed to:

- allocate a stakeholder to each participant;
- appoint an observer of the process;
- appoint a chair from among the stakeholders;
- adopt labels which describe each role;
- hold a committee meeting to establish policy principles/strategies and action plans to manage the issues;
- reach a decision within one hour.

Note: a goal needs to be provided as a focus for the committee's activities, with both a carrot and a stick for reinforcement. The group should be informed that either a $2 billion government grant or a fresh environmental disaster hangs on the outcome of their meeting.

When a decision is reached, everyone 'signs off' from the committee, dealing with any feelings (anger, opposition, argument, etc) which arose during the meeting while still in their stakeholder role. They should then formally leave their stakeholder role, removing their label and speaking from their own positions for a few minutes. The rules by which everyone signs off from a role play exercise are in Appendix I: The Course Coordinator's Guide.

After tea-break, the group reconvenes, and:

- the observers feed back the negotiating techniques, bargaining and alliances they noted;
- members discuss experiences of power and powerlessness;
- members discuss alliances and techniques; and

- the group evaluates the quality of the policy and or strategy that they developed between them.

Committee Procedures: Holding a Meeting

The Purpose of the Meeting

- **To inform:** The easiest to achieve because it is a broadcasting of information. Be careful to keep it interesting, and that all messages are heard.
- **To decide:** The aim is to reach a decision acceptable to the majority of the meeting. The Chair must research the subject before the meeting and know the rules of debate.
- **To make a plan of action:** The aim is to formulate a plan of action following a previous decision or imposed set of circumstances. Suggestions for a draft plan could come from the Chair, a separate committee, or the meeting.
- **To enjoy social contact:** Keep formal business to a minimum but make sure that formal rules have been argued. Make sure visitors are introduced and welcomed. Forestall the development of cliques.

Preparing for the Meeting

Administration

The Chair and Secretary must agree on the purpose of the meeting and work together on the administration and agenda. They can delegate some of the jobs.

When and where. Notice of meeting must give date, time, place and business to be conducted.

Accommodation. Meeting room needs to be prepared. Who has the key? Are there enough chairs? Is there adequate lighting, effective ventilation, cooling or heating? Should smoking be banned? Are toilets available? Is there parking?

Aids. Are all relevant documents available? Will visual aids, such as black or white boards, chalk, felt pens, dusters, easels, slides, or a movie or overhead projector, be needed? Power points and extension cords available? Need for public address system? Who supplies and operates it? A bell or gong is needed call the meeting to order. A timing device is needed.

Refreshments. Will refreshments be served? Who will organise this? Check for heating water and food, crockery, cutlery, tea-towels, serviettes and rubbish disposal.

Agenda

Essential, priorities, timings. Every meeting must have an agenda; it is essential for the Chair and Secretary, and desirable for other participants. It must be realistic with respect to priorities and time. The Chair should ensure that the agenda is acceptable to the meeting. Amend if necessary.

Remember: 'The person who controls the agenda controls the debate'.

Chairing the Meeting

The Chair. The task is to chair the meeting. The occupant of the chair selects 'Chair', 'Chairperson', 'Chairman' or 'Chairwoman' as their preferred title.

Why have a Chair?

To guide, to control. Someone needs to guide the meeting toward achieving its aim. The Chair has the responsibility for controlling the meeting.

Servant-guide, not dictator

Seek agreement, effective debate, will of the meeting. Someone has to ask the basic question – 'Are we agreed?' Someone has to ensure fair and reasonable debate. The Chair is there to ensure that everyone has an equal chance to speak and the meeting makes decisions it needs.

Be prepared ...

Know material, know agenda, speak with the Secretary. The Chair should know the history of the material to be discussed. S/he must have studied the agenda and discussed it and all other arrangements with the Secretary well before the meeting.

The chair

Address the Chair. All speakers should preface their remarks with 'Mister/ Madam Chair' and proceed only when 'noticed' by the Chair.

Control the timing

Realistic maximum time, keep it brief. Allocate time to each agenda item. Trim agenda or extend planned duration. One or two hours is the maximum practical duration of an ordinary meeting. With good planning, most business meetings can achieve their aim in one hour.

Know the rules...

Constitutions, standing orders, rules. Organisations have constitutions, standing orders and rules. The Chair must apply these firmly to retain the confidence of the meeting.

Dissent from the Chair's ruling...

Must be moved immediately after the Chair has given a ruling. The Chair must make many decisions concerning the conduct of the meeting. If a member disagrees with the Chair's decision (ruling) they must immediately announce, 'Mr/Madam Chair, I move dissent from your ruling'. The Chair must immediately explain why the ruling was made. The Chair should then call for a seconder. If there is one, debate should be followed by a vote.

Be no longer heard...

Mr X be no longer heard, moves seconded and debated. Sometimes a member can become very irritating to other members by obstructing the progress of the meeting. Anyone may interrupt and move that Mr/Ms X be no longer heard. If the Chair thinks the motion is reasonable, a seconder is called for, and the motion can be debated.

Out of order

'Mr B, sit down, you are out of order', Chair prerogative. When the Chair considers that a speaker is out of order (such as trying to speak twice to a motion), the Chair should say (standing up if necessary) 'Mr/Ms B please sit down, you are out of order because...' The Chair should be sure of their ground and also be ready for a dissent motion.

Be brave

Take the initiative. To achieve the aim in the allotted time, the Chair should take initiatives such as terminating the debate and putting the motion. Be prepared to use the gavel or the bell to attract attention.

Summarise

Progressive summaries, keep the meeting informed. Progressive summaries by the Chairman are very useful to both the meeting and themselves. The Chair can prevent confusion by periodically explaining where the meeting is on the agenda and by explaining the effect of motions.

Handle

Plan, be firm and fair, have strategies ready. If the meeting is going to discuss contentious issues, the Chair must plan how to deal with conflict of all types. One good way is to state the rules at the beginning of the meeting, and say how everyone will get a fair deal. Distribution of a sheet of rules of debate can help if the meeting is going to be difficult.

Point of order...

Can be raised at any time, Chair must give a ruling. Any member of the meeting may interrupt at any time – 'Mr/Madam Chair – point of order', if they feel that the rules of meeting procedure are not being followed. The Chair must ask the member what is their point of order – give a ruling, and take the necessary option. The Chair is asking for trouble if they ignore points of order.

No confidence motion

Chair should accept with good grace, elect acting Chair. This motion could arise after several unpopular rulings or statements by the Chair. The best procedure is to accept the motion with good grace and if seconded, vacate the Chair and allow debate. After having conducted a quick election of an acting Chair, the original Chair can speak during the debate.

Hypothetical Debate: Saxa Shire

Rob Wiseman

The Story So Far ...*

Saxa Shire is a small irrigation area with the worst soil salinity problem in a bankrupt Australian State. It contains a marginal dried fruit industry, an escalating environmental crisis and an non-viable system of land management. The State Government provides the area with more subsidies than the Shire can return as produce. Despite recession, inflation and the El Niño Effect, the community is still reasonably cohesive. It remains essentially an Anglo-Saxon/Mediterranian farming community although there are an increasing number of hobby farmers from the State capital.

Saxa – the largest town in Saxa Shire – is playing host to two groups of environmental experts. The Shire President has seized upon the opportunity, and invited them to a meeting of Council and community leaders to determine directions for Saxa's development.

The first of the visitors is the new State Minister responsible for the Environment. The Minister is visiting the region to ascertain for herself the damage that has been done to the region. She is looking to find solutions to the physical, biological, social and economic crises that face the Shire. Accompanying her is a small team, including her Departmental Secretary and main political adviser.

---

* For further background see Chapter 7: *Saving a shire: a salt-affected irrigation district in New South Wales.*

Environmental issues have not been high on the agenda for the present State Government up until about three weeks ago, when they were perceived as an election issue, and a new Minister was appointed. The Minister's inaugural speech on the steps of the Department was widely reported in the press, and has been hailed as a sign of changing attitudes within the government.

The Department of the Environment has languished for many years as a government backwater – a view held especially by those high in its administration. The Department has precious few resources and even less authority, and runs on the philosophy of community self-help rather than State intervention – a role enforced as much by economic necessity as enlightened policy. Nonetheless, the Department has received increased interest lately, and will capitalise on that to improve its position. A solution to the 'Saxa Situation' would be a feather in its bureaucratic cap – especially if it can do it at no cost to itself.

The other group in town is a scientific research unit, here to investigate environmental and social change in Saxa Shire. Based in the National Capital, they are part of a federally funded research programme investigating social, economic and environmental problems around the country. They first visited Saxa six months ago, and their first report painted a gloomy picture of heavily salinated soil, steadily decreasing farm productivity, and increasing social stagnation. They also pointed out that all these problems could be attributed to the small size of farm blocks in the area – a legacy of nineteenth century planning – and the intensive farming techniques used in an attempt to make them profitable. Unlike several previous reports on the state of the Shire, the community has reacted strongly, seeking answers to the problems revealed by the study.

The researchers have returned to Saxa to complete their study and see how the community is dealing with the situation. They are eager to be of assistance, and are willing to contribute their vast experience to anyone prepared to listen to them.

The Shire itself is economically depressed, socially stagnating, ecologically tottering and politically conservative. Residents are keen to avoid change, but equally keen to see things get better – an attitude that has its source in nearly a century of government interference. Recent council elections returned all councillors with one exception (the incumbent moved interstate). The new councillor has proved to be an active stirrer. Keen to get things done, she remains somewhat unsure of the internal workings of the Council. This has not noticeably dampened her enthusiasm for 'community government' as she calls it. The Shire President and Clerk do not believe that she is anything more than a passing annoyance, and are prepared to tolerate her.

The Shire contains several other important individuals and organisations who have a hand in the environmental and social change within the Shire. The Land Conservation Officer and the Shire Engineer are responsible for environmental planning in Saxa. They have never seen eye-to-eye and have vastly differing views of what needs to be done. Both believe that they are the sole experts on Saxa Shire's environment.

In the community there is the Women's Group – an uneasy alliance of the CWA, Single-Mothers Collective and the Women's Volunteer Service – the small but growing body of hobby-farmers from the State capital, and the local Fruit Growers Association which represents the mostly Mediterranean fruit-growers and small landholders.

The Shire President has convened a special meeting in the Council Chambers to bring together the community and plot a course of action to save Saxa Shire. He is hoping to get funding from the Minister (whom he has invited to attend). Also present will be community leaders including the bank manager, a cannery owner, the Shire's largest landholder and representatives of the major community groups. The President has also asked the scientific team that first showed the extent of Saxa's environmental degradation to contribute their experience.

The characters

- **The Minister**: has only been promoted to her post a few weeks ago on the promise of 'equitable and economical environmental management' – whatever that is. She is keen to have an early success in her new portfolio, and Saxa Shire is her first big challenge. Although ostensibly on a fact-finding visit, she would like the issue settled while she is in town – presumably so that she can claim all the credit. A keen and enthusiastic politician with a large majority in her suburban electorate, she is willing to support almost anything, so long as it buys her publicity and popularity.
- **The Departmental Secretary**: desperately trying to keep his new Minister out of trouble and on established departmental policy. He is the only one fully aware of the extent of funding shortages, and is the person responsible for much of his Department's expenditure. With this in mind, he is trying to stop the Minister from 'doing anything rash' while in Saxa. Although not opposed to change, he anticipates the added strain it will put on his already overburdened staff. This man is no relation of Sir Humphrey* – although he occasionally wishes that he had the same influence over his Minister.

---

* *Yes, Minister*, BBC Television

- **The Minister's Political Adviser**: the party man whose occupation is to keep the Minister out of hot water and get the Government re-elected. This is a role that sometimes coincides with the departmental secretary's objectives – but often as not, opposes them. He is well aware that good politics and good government are not always compatible, and will normally opt for the former if pressed. This man has the Minister's ear, and like most party hacks is always on the lookout for 'party contributions'. Keen to score himself a cushy job in State Parliament, any success for the Minister is a success for him.

- **The Shire President**: just re-elected for another term (his fourth) on the basis of a strong personal following rather than any policy agenda. His attitude is 'We know what we're doing, and we don't need anyone to tell us anything different'. Or at least that was his attitude until he saw a scientific report six months ago, showing the expected decline in 'his' Shire's productive capacity over the next ten years. Consequently, he is in a state of flux between 'the old ways' and the promise of an improved future. Mistrustful of the Minister (she is a suburbanite from the wrong party) but is willing to cooperate to get the funds he needs to solve Saxa's problems.

- **The Shire Clerk**: would like to think of herself as the power behind the throne – better yet, on the throne. She sees the Shire President as an inconvenient hindrance to her authority. She has been the Shire Clerk for nearly twenty years now. She has nostalgia for the Saxa that she grew up in, and is consequently mistrustful of change: as she sees it, change is what hurt the Shire in the first place. Although she is concerned about the social changes in Saxa, she is uncertain of what to do. She controls the Shire administration, and is keen to see that her position is not eroded.

- **The Land Conservation Officer**: banished to Saxa (her view) for the last five years now. She is presently halfway through her doctorate and is keen to return to the State capital (and her fiancé) to complete it. Despite a brilliant academic background and her long presence, she has had only a marginal effect in changing irrigation techniques and land use, and halting the spread of soil salinity. This is not entirely her own fault.

  She is well liked and respected in the community, but feels threatened by the presence of the scientific team, and sees it as something of a slur on her ability. She wants to save Saxa and salvage her professional reputation but is torn between the demands of departmental policy and professional pride.

223

- **The Shire Engineer:** the main reason that the Land Conservation Officer has been so dramatically unable to change the Saxa situation in the past five years. A strong supporter of the *status quo* for as long as he has been in Saxa – nearly thirty years now. A traditionalist at heart, he believes that the best solution to a problem is an engineering solution, and is deeply resentful of anyone who suggests otherwise. Very interested in new technology on the grounds that it will allow him to build bigger and better dams and irrigation channels. Although not impervious to the situation in Saxa, he sees it as someone else's problem.

- **The New Councillor:** elected on the platform of 'Time to Change'. She is well in touch with the local residents and their concerns, and is keen to find a long-term solution to the 'Saxa Situation'. She is rather less in touch with the operation of the Council and has put several noses out of place – notably the Shire Engineer's. She has her eye on the Shire President's job, and would like to think of herself as ruthless and efficient (although ambitious but compassionate would be better descriptions).

- **The Bank Manager:** opportunistic and looking for a big city bank to manage. Most (but not all) considerations come secondary to this. She manages the only major bank in the district. Most of the Shire has either debts or mortgages with her bank, and sees any change in the Shire's structure as inevitably costing it dear. In these depressed (recessed?) economic times, she is working hard to keep her books in the black. Almost certainly immune to being bribed.

- **The Large Property Owner:** Although painted as one of the 'white-shoe brigade' by many residents, she is actually quite concerned about the state of Saxa Shire – and not for entirely economic reasons either. She has a private nature reservation in the area and has contributed to the support of local revegetation and soil conservation projects. She owns something like 30 per cent of the Shire – with an even split between irrigation and grazing. Her properties and farms employ many Saxa residents. Her holdings here are only one of a number of properties and investments she has around the country; they are also the least profitable. Although she would like to see an increase in profitability, she is resigned to the fact that she may have to write off her assets if advised to do so by her financial experts. She has connections with the Minister and has made several large contributions to the party's coffers over the last decade. Also a generous contributor to many community and welfare groups.

- **Cannery Owner:** a local boy made good. From a local Mediterranean family, he has supported the community through the construction of a cannery – which incidentally employs many of Saxa's young people. Keen to keep the cannery open, but financial pressures are building up. His full contribution to the community is not fully recognised. He is in an agony of indecision over whether to move the cannery to more profitable areas, or to stay unrecognised with the community he has lived in all his life.

- **The Women's Representative**: a comparatively recent arrival, she has forged an uneasy alliance of the CWA, Single Mother's Collective and the Women's Volunteer Service (an important group in the community). Highly, if not overly, idealistic, she feels that women have a major role to play in the restoration of Saxa – especially as they play such a large role in the service and support sectors. What exactly she has in mind is not so clear. She is involved in just about every community event, and has organised the local Landcare group. She is aware of the problems facing country women, and wants to draw attention to their needs. A busy and active member of the community, and if she is regarded as something of an oddball, she is generally accepted. A great believer in equality and having her say.
- **The Farmers' Representative**: represents a mostly Mediterranean constituency of tenant fruit growers. This group has watched their crop returns go steadily down over the last twenty years. Also a part of the group are several graziers, mostly raising sheep on unirrigated land. Although the farmers and growers are not a wealthy sector of the community they do produce a fair proportion of the Shire's earnings. They are a mostly aging population with a large debt burden, concerned about the loss of productive land to salt, banks and hobby farmers. They are keen to be involved, but not sure what to do.
- **The Hobby Farmers' Representative**: representing a much maligned but growing sector of the population. Mostly refugees from the State Capital looking for rural peace and a small plot of land, they are increasingly taking over productive agricultural land. Although the blocks they are occupying are not large enough to support extensive crop or fruit growing, local farmers are concerned about land being tied up in this way. Another concern is that most hobby farmers are inexperienced, and there has been a high turnover of farms as disillusionment has set in. Hobby farmers represent a wealthier sector of the community. It is a matter of some resentment that their affluence does not flow out into the community.
- **The Scientists**: a group of eminent and highly energetic – although breathtakingly innocent – scientists looking into the problems facing Saxa Shire. They are tolerated by council and local community on the

grounds that they may attract attention to the plight of the Shire. They might even propose some solutions, but no-one is hoping for quite that much. They have access to vast amounts of technical and social data, which could be of great value to the Shire. They are trying to play a straight bat, and help everyone, but their funding is running out shortly and they are looking for some form of support for their work.

- **The Moderator**: Do not expect Geoffrey Robertson. The Moderator is present simply to keep the story moving as necessary and to provide the occasional ruling. The Moderator's word is law, and is the only one present guaranteed to be unbiased and unbribable.

History

Saxa Shire was first settled following the market crashes of the 1880s. The initial concept was to develop a cooperative village filled with part-time farmers. Blocks of land were offered that were too small to be viable independently, but sufficient to provide against the unemployment being experienced at the time. The aim was to avoid a community of full-time farmers that would be susceptible to varying economic fortunes. The backers of the plan were aiming to develop a cooperative community. The area was to be irrigated – as much of the nearby riverland already was – to support vineyards and orchards.

Despite initial enthusiasm, the necessary support, in the form of pumps and finances, did not eventuate. Dreams of a lush irrigated region producing grapes and stone-fruits were replaced by a reality of small scale hay and lucerne production. As the original planners had designed, the small properties were too small to be able to support a farm, especially without an adequate water supply.

Soldier settlements established in the area after the Great War quickly failed, through a combination of inexperienced farmers and a lack of support. The irrigation system remained inadequate, a railway facility was not fully developed, and Saxa Shire was cast into competition with neighbouring, more established areas. To make the situation even more difficult, neighbouring Shires vigorously opposed the development of Saxa Shire, seeing it as a potential threat to their own livelihoods.

By the 1920s, land degradation was becoming a serious problem, partly because the soil types that occur in Saxa Shire, and partly through inefficient farming and irrigation methods. As the area of salt-affected soil grew larger, farmers began to walk off the land. A dried-fruit industry did become established in the region, but it lacked the size or the profitability of other irrigated areas. There was little incentive for farmers to change crops – during profitable times, it was economically unattractive, and in lean times, farmers lacked the capital to afford the change.

By the 1940s, much of the low-lying area had been abandoned to the salt. However, the upsurge in prices during the war years meant that the region was able to afford to build an adequate drainage system. Plans were drawn up for tile drainage throughout the region. The plans were never implemented – partly because of the lack of manpower and materials during the war, and partly because of arguments over exactly where drainage ditches should be placed. A series of bad seasons and poor markets after the war effectively ended the optimism and opportunity for change.

By the 1950s, the State Government had come to the realisation that Saxa Shire did not have a suitable climate for a dried fruit industry, a problem compounded by poor soil types and increasing salinity. The block sizes were judged too small for a farmer to generate an acceptable income. Many blocks were saddled with significant debts. The recommendation of an enquiry into Saxa Shire was that blocks be amalgamated to increase their size, and crops diversified to provide a more secure economic base for the community. The government of the day chose to address only the second, politically-easier recommendation.

However, different crop types posed their own problems. Most stone-fruits simply could not grow in the Saxa soil, and there were no canning or packing facilities located within Saxa at that time. Prunes and apricots were a better alternative since they could be dried as grapes already were. Vegetables were a more promising option, but they required more water than the Water Commission was supplying.

Not surprisingly, neither of the recommendations were popular locally, and were resisted, with farmers and growers preferring to stick to what they knew and cope with the future as best they could. Meanwhile, although a drainage system had been constructed by the Water Commission, many farmers were not connected to it, despite clear salt and erosion problems. They simply could not afford to.

By the 1970s most of the irrigated land was committed to vineyards – chiefly sultana grapes – although there were some growers who had diversified to stone-fruits and vegetables. New irrigation techniques such as sprinkler systems and microdrips had been introduced, but over 90 per cent of irrigation was still done by the traditional furrow and flood methods. There were several advantages in the new systems: larger areas could be irrigated, as smaller amounts of water would be needed, and the effects on the water table would likewise be smaller.

In the mid-1970s the 'Vine Pull Scheme' was initiated with the aim of removing old vines and getting farmers to take on new crops. Although many vines were removed, new crops did not eventuate. Some farmers used the financial incentives offered to replace old vines with new, others simply removed the vines and left – leaving the ground fallow. This situation was the beginning of the break-up of the Saxa community. There has been a rapid turnover of farms since then – mostly to hobby farmers. The district was hit hard by drought in the early 1980s. Ecologically and socially, the Shire has not yet recovered. Although the community boasts as a wide selection of shops, clubs, sporting facilities and social activities, as it did fifteen years ago, there are signs that the area is not as vibrant as it was. However, residents are determined and far better prepared than before to cope with the problems that the Shire faces.

Physical Conditions

At the moment, approximately two-thirds of Saxa Shire is given over to irrigated crops – mostly vine-grapes and vegetables, with some stone-fruits and citrus – and irrigated pasture. There is a small amount of specialty farming; chiefly nuts and avocados. The remainder of the agricultural land in the Shire is used for sheep grazing.

The majority (95 per cent) of irrigated properties are under 20 hectares – the average size is only nine. These blocks are too small to be viable individually. Unfortunately, present planning by-laws make subdivision or amalgamation of blocks difficult. Most of the low-lying areas are badly affected by salt, especially land that has been cleared of native vegetation for irrigation. Erosion is another serious form of soil degradation in the area.

Although areas of native forest and grasslands within the Shire have been preserved as national park, animal and plant species have been disappearing – mostly as a result of rising salt-rich water tables.

Social Background

The traditional farming community comes from a mix of Anglo-Saxon and Mediterranean backgrounds. There have been some recent arrivals – mostly hobby farmers from the State capital – that have begun to change the social mix of the area.

The population is ageing, with over 40 per cent of the population over the age of forty. The population level is approximately static. The position for young people is not promising. There is no local high school, and students have to be sent to a neighbouring township for schooling. As in many other rural areas, youth unemployment is high, and job opportunities are low. Many young people leave for larger population centres if they are unable to find work locally.

As the population ages, many farmers are opting to stay on the land rather than move into town – with the result that increasingly large parts of the Shire is becoming non-productive. Inexperienced and inefficient hobby farmers, attracted to the region by the dream of rural peace, are using precious water and soil resources to produce, in agricultural terms, far less than the growers they replace. There has been a high turnover of farms – mostly as new arrivals grow disillusioned and leave, to be replaced by fresh hopefuls.

One particularly divisive community issue is that of who pays for the provision of irrigation water and channels if farms are not actually using these resources.

The town of Saxa is like hundreds of others in rural Australia. It boasts a wide range of services, shops, and amenities. However, most were set up in past decades, not recently, and the town is beginning to stagnate. Nevertheless, a sense of community still prevails, and people are prepared to change and be involved in change if it means a better future for Saxa.

Women organise many of the community's services such as Meals on Wheels and the Landcare group. While their contribution to the community is recognised, it is usually undervalued. As a group, women do far more than any other sector of Saxa Shire to promote and maintain community cohesion.

Community services include:

- one major bank with two smaller branches,
- three primary schools, and numerous facilities for young people;
- a hospital, nursing home, Day-Care Centre, ambulance service and chemist;
- several light industries – although the major employer is the cannery;
- a wide range of shops and produce outlets;
- a lawyer, solicitor, doctor and several churches
- sporting facilities – including race track, trotting complex, lawn-bowls, football and cricket clubs, tennis courts, golf courses, and a shooting range;
- many clubs – including the RSL and CWA;
- many social and support organisations including Meals on Wheels, Apex Club, Lions Club, Senior Citizens, Women's League, Red Cross Society, Volunteer Bushfire brigades, Saxa Youth Help Group, the Single Women's Collective, and the Women's Volunteers Service.

Local Economy

The major economic activity of Saxa Shire comes from the dried fruit industry which produces mostly sultanas, currants, and dried prunes and dried apricots. Some of this produce is marketed by individual farmers, and the rest through a branch of the Farmer's Collective. Vegetables reach the markets in much the same manner, although they do not produce as great a financial return.

The other main source of income in the town is the growing and canning of stone-fruits. The cannery opened approximately ten years ago, and now employs approximately nearly fifty people, mostly young. However, decreasing crop sizes and deteriorating fruit quality have combined with a poor economic climate to bring the cannery's viability into question.

Some lucerne and hay is grown; some is sold outside the area, but most is used locally. It is a relatively low value crop, and supports only a few individuals.

There are many local industries – vineyards, packers, grocers, butchers, hardware, garages, nurseries and builders – but they largely support the local community rather than bring in outside funds.

In the past, there have been more industries in Saxa, and the Shire generally, but most have drifted away; drawn by better opportunities in other areas. This has also had the effect of moving the centres of employment away from the local community, so that some people are forced to move or travel long distances each day to earn a living.

The other major source of financial support for the Shire comes from the State Government in the form of subsidies and funding for capital works. A significant proportion of the population receives some kind of social security. A recent analysis carried out by the Department of Finance revealed that the value of subsidies given to the Shire is greater than the value of the crops grown in it. This is not widely accepted in the Shire, but readily believed in the State Government – where bankruptcy, recession and economic rationalism have ensured that cost cutting is at the top of the political agenda. The Minister is under pressure to cut spending in her portfolio, but she has a promise of 'equitable and economical environmental management' to implement (or at least explain).

The Process

The Moderator will start the proceedings by inviting the Shire President to welcome those gathered to Saxa Shire, and to outline briefly the history and nature of the difficulties facing it – perhaps delivering a few home truths about their situation. The Minister has agreed to outline how her new environmental policy will benefit the area, and perhaps even offer some kind of support to Saxa. The scientists will provide a short overview of the physical and social state of the Shire. Others present will then be free to state their positions, and the issues that they see as important in the debate.

There will then be a short break of fifteen minutes, so that all parties may mix informally, bargain furiously and establish grounds for negotiations; and to allow for refreshments.

The Conference will then reconvene, and the meeting discuss the issues facing Saxa Shire. Smaller interest groups may form to discuss particular issues. Any plans developed must have the support of all those directly involved. Discussion and breaks will continue as necessary until a position is reached or talks break down. The issues facing Saxa Shire need to be clearly identified and solutions to them found. Although not every individual needs to contribute to the solution of every problem, exclusion of interests could jeopardise future plans.

The Moderator will assess the effectiveness of the final strategy on the basis of economic and ecological sustainability, social equity, levels of community, Council and Government support, and the likelihood of a successful outcome. These issues need to be addressed by the meeting. Some issues will be more important than others, and some individuals will not need to contribute to all solutions. However, beware of excluding interests in any plan. Although these issues have been presented separately, in Saxa, they are all interrelated.

A Final Word

Although the issues presented in this hypothetical case are very serious, the approach to solving them is intended to be rather more lighthearted. Be prepared for greed, corruption, bribery and treachery to appear at any time. No-one is entirely reliable, and everyone has their own barrow to push.

The character portraits that have been provided give only a broad description of each individual's background and motives. Furthermore, they are only the things that everybody knows about – there may be deeper and more sinister motives floating around. Be prepared to extemporise. The essential thing is to get into the part of your character – no matter how greatly you despise their attitudes or motives (the correct selection of liquid refreshments has been known to help this process along splendidly!). Above all, have fun while you stab people in the back, lie through your teeth and do deals. You can even try dealing openly and honestly if this appeals to you. Remember, the aim is to solve the problems and enjoy yourself.

This hypothetical study is based on a very real situation – Nyah Shire in Victoria (Chapter 7), although a certain amount of elaboration has been necessary for the sake of the story. The Saxa Shire Case Study can be used as a source of further information, but this is not absolutely necessary: the background information and character sketches should be quite sufficient. If anything is missing, invent it or ask the Moderator.

The characters are entirely fictional, and bear no relation to anyone – living, dead or in politics. Any resemblance is purely coincidental.

13 Stage 3: Goalsetting and Negotiation

All the interests in the solution

LEARNING MATERIALS:

- preparing for negotiations
- three party coalition exercise
- a negotiation exercise: Irrigation Inc

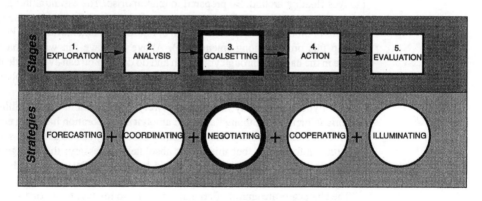

Figure 13.1   All the interests in the solution

AIMS AND OBJECTIVES

This learning session is intended to:

- provide experience in negotiating on behalf of an interest group;
- provide experience in negotiating to mutual advantage, rather than win–lose or win–win;
- identify others' negotiating styles; and
- extend each participant's own negotiating skills.

This session makes use of detailed case studies of the negotiating positions on one environmental issue. While this has some problems for those who usually avoid simulation exercises, it is a safe way to practise negotiating skills. To implement the skill of negotiating to mutual advantage back in the workplace, without some practice, means risking losing to the traditional competitive win–lose strategist.

The Harvard Negotiating Process recommends four steps for negotiators who seek to go past a win–lose result to an outcome which is better than any of the players could achieve alone.

The four steps are:

1 Separate the people from the problem: we did this in Stage 1.
2 Focus on interests, not positions: we did this in Stage 2.
3 Invent options for mutual gain: this is the challenge for this session.
4 Insist on objective criteria: this will be the subject of Stage 4.

There is a choice of four case studies in Section 2, each of which supplies the content for environmental negotiations. These case studies can be used by parallel teams, each taking one set of issues and comparing the results at the end. They could be used by all participants in sequence throughout a longer programme, or one case study could be selected as the most suitable for a particular group.

This chapter concludes with a simulation game, developed to illustrate cooperation and compromise, based on the case study of Calico Creek.

RISKS AND OPPORTUNITIES

Stage 3 of managing environmental change requires negotiating to mutual advantage between all the key stakeholders. This means going beyond the zero-sum game of 'if-I-have-more-you-get-less' to another way of thinking. Since there can be no winner without a loser, the notion of winning at all becomes a danger to constructive negotiations. Victory for any one party is defeat for another. The outcome could be to *everyone's* mutual advantage.

233

Advice on conducting principle negotiations from Fisher and Ury's book *Getting to Yes* includes:

- changing the game;
- negotiating on the merits of the case (not on your or others' predetermined positions);
- treating the game as mutual problem-solving;
- being soft on the people, but hard on the problem;
- exploring the interests of all players;
- avoiding having a hard bottom line;
- developing multiple potential options;
- leaving decisions until later when all the evidence is in;
- reasoning and staying open to reason; and
- yielding to principle but not to pressure.

The BATNA

A valuable concept from *Getting to Yes* is the idea of the BATNA – Best Alternative To a Negotiated Agreement. If you have decided on this early in the negotiations, it prevents you from becoming locked into fruitless negotiations. The idea of the BATNA is developed in *Getting to Yes*.

ACTIVITIES

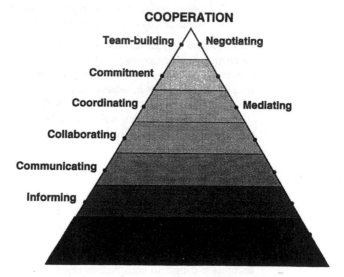

Figure 13.2 All the interests in the solution: activities

Theme: Commitment to a Shared Goal

Choosing any of the issues in the case studies of Section 2, participants will need to consider the issues in some depth, building on the information gained in the previous three learning modules. The best mode of approach is to continue with the issue selected above. The negotiation process will involve taking part in a simulation exercise that will include:

- **Communication:** identifying the interests involved;
- **Commitment:** inventing options for mutual gain;
- **Arbitrating:** arriving at agreed, objective criteria for a solution;
- **Team-building:** debriefing from the exercise roles.

The exercises build on the previous three learning modules and allow practice in the following:

- **Negotiating:** negotiating to mutual advantage;
- **Mediating:** improving on expected outcomes;
- **Arbitrating:** competence evaluation; and
- **Communicating:** learning from experience.

Each of the case studies can be used in its own right as the context for practising negotiations. Each negotiation should take three hours for maximum learning.

For Nyah Shire, there is a *hypothetical* debate which allows participants to explore different options for the solution of the issues, without being tied to the results (see Chapter 12).

For Calico Creek there is a *structured negotiation* which allows practice in cooperation between key stakeholders where the rules are fixed beforehand (see Chapter 13).

For Bangkok, there is an exercise duplicating an *open forum*, which would be one method for achieving commitment to a common goal between apparently disparate interests (see Chapter 14). This is an appropriate method for a meeting between community, developers and administrators.

Before embarking on the more in-depth case studies, everyone should read the notes and undertake the 'three-party coalition' exercise below.

PREPARING FOR NEGOTIATIONS

Ian McAuley, Consultant, 24 Schlich Street, Yarralumla, Canberra, Australia*

*   From notes developed by Professor Max Bazerman, and from the ideas of people involved in the Harvard Negotiation Project, especially Roger Fisher, David Kuechle and Howard Raiffa.

1   **What is my best alternative to a negotiated agreement (BATNA)?**
The answer is determined by the use of power in the negotiation. I am far stronger in a negotiation when I am willing to walk away. Even if I have little inclination to walk away, a good BATNA will increase my confidence.

2   **Comparatively, how important is each negotiation issue to me?**
I should know the answer before the negotiation begins. I can compare the attractiveness of alternative agreements and create mutually desirable trade-offs. A scoring system is one way to create this preparation.

3   **Can I articulate my interests?**
Have I really got my interests clear? Can I articulate them in a way that the other side will understand clearly?

4   **What is the other side's BATNA?**
Negotiators often underemphasise the importance of this assessment. However it gives my best estimate of how far the other side can be pushed. It also keeps me from making offers that lead the other side to walk away. A good negotiator will not disclose his or her BATNA – my best preparation may be to try to put myself into the other party's shoes.

5   **Comparatively, how important is each issue to the other side?**
This information helps me develop trade-offs. It also gives me information on which issues are central in the minds of the other party. Interestingly, the other side will (and should) often be willing to provide this information.

6   **What do I know about their interests?**
It will help if I research this in advance, but I should not get locked into my own perception. I should be prepared to listen during the negotiation. If the other side is unlikely to state their interests, I should be able to say 'Is it correct that your interests are ...?'.

7   **Who are the negotiators?**
What can I find out about them? Can I establish rapport before the negotiation? I should beware of others' stereotypes ('She's a tough cookie.') I should beware of my assumptions – I already know the person in another role (friend, work colleague), but how might they behave in this negotiation?

8   **What are the norms of the negotiators and those they represent**
Compare expectations of norms of behaviour (eg with regard to truthfulness) in negotiations between, for example: a) spouses in a

good marriage vs spouses in a troubled marriage; b) businesses with an ongoing relationship vs businesses in a once-off situation.

9   **What are the facts, options, trade-offs?**

I will be stronger in negotiation if I have already done my technical homework – researched key facts, made calculations, developed financial models etc. The negotiation table is not the ideal place to get involved in complex calculations or to have to call for factual material.

10   **Have I attended to all details?**

Where do we negotiate? (This does not apply only to large formal negotiations – do I negotiate with my bank manager at the counter, in her office, or over lunch?) How do I present myself? Where do we sit – Party A facing Party B or intermingled? Where do I place myself in relationship to my adversaries and allies?

11   **Are there more than two parties?**

There are important differences between two-party and multi-party negotiations. For example, with more than two parties, coalitions may form and act in concert against other parties.

12   **Are the parties monolithic?**

Rather than being the exception, it is probably the rule that each party to a dispute is not internally monolithic. Where there is not one person on a side who can make decisions, there must be internal as well as external negotiations. When I negotiate a brief from my own party, I should make sure I have a wide brief, with a delegation to negotiate within my side's *principles*. I should avoid being sent as a representative to put fixed *positions*. I should beware of raised expectations on my side.

13   **Will negotiations be repetitive?**

Where bargaining is repetitive, each disputant must be particularly concerned about reputation and the long-term relationship. That may lead to more integrative negotiations than in one-shot negotiations. Note, however, that it does not necessarily follow that those engaged in repetitive negotiations will adopt a 'problem-solving' or 'cooperative' attitude. A negotiator may want to establish a long-term reputation for toughness, and thus adopt a posture which might be seen as counter-productive if viewed only in the short-term, single negotiation context.

14   **Are there linkage effects?**

Where one negotiation is linked to another, the calculation of costs and benefits of a particular agreement is likely to be affected by the implications for the other negotiations. Linkages can create impasses, as, for example, where a party facing multiple lawsuits with different parties over the same issues may not be able to offer a settlement in a particular suit since the single settlement may lead to different, and costly, settlements in other suits.

15 **Is agreement required?**

Must both parties reach agreement? Can one party or both parties walk away, either entirely or at a particular state in the negotiation? This affects the importance of developing an alternative to a negotiated agreement.

16 **Is 'ratification' required?**

The need (actual or feigned) for ratification, as by a corporate board, of a negotiated resolution often provides a vehicle for negotiation game playing, as when one party seeks to squeeze one more concession from the other side during the ratification process.

17 **Are there time constraints or other time-related costs?**

When the North Vietnamese came to Paris to negotiate an end to the Vietnam War, they rented a house on a two-year lease, and let that fact become known. The party who is in a hurry (and lets that be known) is at a disadvantage.

18 **Have we allowed adequate time for negotiation?**

Negotiation is not 'one minute management'. Negotiation often involves learning (and un-learning), accepting new realities, passing through different phases of tension and relaxation. It may be necessary to give time for recesses, to allow negotiators to develop wider briefs from their own constituents. Breaks also help in making transitions between phases of a negotiation – eg from exploration of interests to problem solving.

19 **Are contracts binding?**

Negotiation strategy and arguments on terms of an agreement should be influenced by whether and how the agreement should be enforced. Both sides should agree on how flexibly the agreement is to be interpreted. Be beware of the 'I thought it meant ...' risk.

20 **Who should make the first concrete offer or demand?**

Consider the problem of a) not being so extreme as to destroy the ambience of the negotiation and b) not being so conservative that your offer or demand falls well within the other party's acceptance region. What are the advantages and disadvantages of making the first offer or demand? One should not let an extreme first position be a reference point for further negotiations. If possible bring in a figure which doesn't anchor too rigidly ('How much did the last house in this area sell for?')

21 **How should propositions be put?**

One should try to put 'yes-able' propositions. Can you reframe the proposition in a way which is acceptable to both sides, and to those to whom you are accountable?

22 **How should the negotiation end?**

Don't gloat. 'A final word of advice', offers Howard Raiffa, 'don't gloat about how well you have done. After settling a merger for $7 million, don't tell your future partners that your reservation price was only $4 million: that won't make them feel good. You might be tempted to lie for their benefit and make a vague claim to a reservation price of about $6.5 million. But lies, even beneficial ones, generate their own complications. Some confidential information should remain confidential even after the fact'.

Three Party Coalition Exercise (30–90 mins)

Participants form groups of three. The exercise can be repeated with each person rotating the rules of groups A, B and C.

You are the negotiator for:

- Group A _____
- Group B _____
- Group C _____

Groups A, B, and C are three independent organisations. Each has designated a representative to send to a three-way negotiation. The representatives are empowered to commit their organisations. The three groups have been told by a higher authority that if they work together there are benefits to be had. Indeed, the available benefits are quite explicit.

If A, B and C work together, they can share benefits totalling 121. How they want to divide up the benefits is up to them, but they have to agree upon the exact allocation of benefits before they will be made available. If only two of the parties work together, there are lesser amounts of benefits available (see schedule below). Again, any pair that decides to work together must provide an explicit allocation plan before the benefits will be handed over.

Only one agreement is possible. That is, either the parties agree to a three-way allocation or two of the parties decide to work together leaving the third party with nothing.

Table 13.1 Schedule of benefits

A alone	0
B alone	0
C alone	0
Just A and B together	118
Just A and C together	84
Just B and C together	50
A, B and C together	121

Each representative's goal as he or she enters these negotiations is to get the greatest number of points in the allocated time.

The three representatives should meet together initially to introduce themselves and start the negotiations formally. Once negotiations begin, you will have about 40 minutes to try to reach an agreement. If two of the three representatives wish to speak privately, the other representative may not interrupt for four minutes (although he or she may listen to what the others are saying). If any agreement is reached, it must last for at least five minutes before negotiations can conclude. Two of the three representative can conclude the negotiations.

Results		
Agreement	*Yes*	*No*
Scores	A _____	B _____	C_____
Discussion		

A, B and C should share the experience of the negotiation with each other. Each member takes five minutes to describe their decision-making processes, their learning stages and their personal feelings. The exercise provides experience in logic, interpretation of other's actions and self control!

Negotiation Exercise: Irrigation Inc

Rob Wiseman

Introduction

Irrigation Incorporated is a new company formed to develop and operate irrigation areas in the State. It is interested in the development of smaller, more specialised irrigation areas than have previously been attempted. The company has been formed through the amalgamation of several smaller interests with experience in irrigation, together with several leading financial institutions. The member organisations all have experience with different aspects of irrigation development and management. Irrigation Inc has the capital to fund its projects internally. After several studies, the company has identified Trickle Creek as an ideal site for its first project. Several preliminary discussions have taken place and plans drawn up, but the project cannot proceed without the permission of the State Water Commission (SWC).

For additional background refer to Chapter 9, *The Future of Calico Creek: local management of a small subtropical water catchment.*

Trickle Creek

Trickle Creek is a small agricultural community located in southern Queensland (see Figure 13.3). The region produces small crops such as beans, squash, zucchinis, tomatoes and cucumbers, tree and vine crops such as paw-paw, avocados and passionfruit, together with dairy cattle and a few beef cattle.

The water in Trickle Creek is used mostly to irrigate crops and pasture, with a small amount going to water stock and provide domestic supply. The water is untreated and unpolluted, and the supply is usually abundant in all seasons in most years. The area is subject to drought, however, and most landholders have several dams or weirs as security against this. The neighbouring catchment of Meatloaf Creek has an irrigation scheme – constructed about fifteen years ago – which also supplies the lower parts of Trickle Creek. Trickle Creek farmers above the regulated supply have to rely on natural runoff, supplemented by bores, dams and tank-water.

The headwaters of Trickle Creek rise in what the locals call the 'slip country' – a reference to the landslide prone region north of the farming community. This catchment also contains a small State Forest reserve. (For more detailed information on Trickle Creek, see the previous section).

The Project

Irrigation Inc proposes the development of an irrigation project to service the farming community of Trickle Creek. The small size of Trickle Creek and the nature of the 'micro-irrigation technology' developed by Irrigation Inc make this project unique in Australia. If successful, the new techniques Irrigation Inc has developed will allow the development of a significant number of other marginal rural areas.

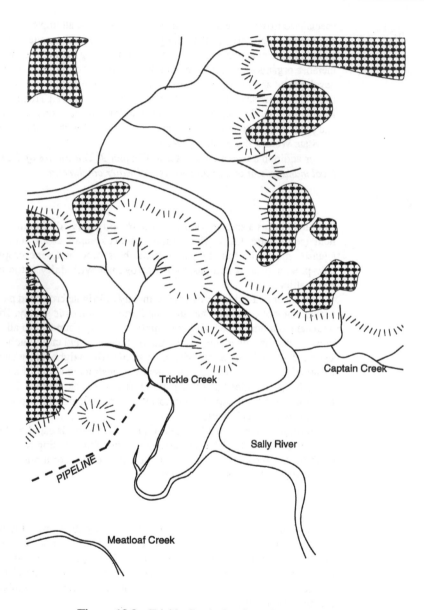

Figure 13.3 Trickle Creek, Southern Queensland

The heart of the project is the development of two dams in what the locals refer to as the 'slip country', as well as the edges of the State Forest reserve. The dam sites themselves are not utilised as they are too steep to be farmed. The catchment will be cleared and the dam sites deepened, with the excavated earth to be used in the construction of the dam walls. The creek channel will be rerouted as necessary to deliver the water closer to the areas it is needed, so as to reduce the amount of pumping necessary. Canals and water channels will also be dug for those irrigators prepared to pay for them.

The project construction could be expected to be completed in about 18 months from commencing work, although full storage and water quality facilities may take up to five years to be fully operational. The overall cost of the project is estimated at $10 million.

The Approval Process

Irrigation Inc submitted its plan to the State Water Commission for approval six weeks ago. Given the 'obvious benefits to all parties', Irrigation Inc saw little difficulty in obtaining approval for its application. It hopes to begin work in Trickle Creek in six months.

The SWC has been under close scrutiny from State Parliament, however, for its handling of several previous State Water Projects, especially in relation to the allocation of funds and the quality of 'environmental management'. Consequently it is particularly sensitive to the level of support such a proposal receives, and the amount of environmental and social disturbance the project would create.

In this particular case, the Commission will not approve Irrigation Inc's application unless there is clear support for the project to proceed. It has decided that, of the six parties that it has identified as 'key players' in this issue, at least five of them must support the proposal. (The Commission would like to see unanimity if possible, but five out of six players will indicate a sufficiently strong level of support).

Two parties or more can block the proposal. Irrigation Inc can also veto any proposal as it is the only one with the capital and experience to develop Trickle Creek.

The Players

Irrigation Inc is excited at the possibilities of the Trickle Creek project. It believes that the project is not only viable but profitable – both for itself and for the farmers on the scheme. Its enthusiasm is based on data compiled by the State Water Commission, and its own advanced computer simulations. There are also longer term advantages as well. If the project is a success, it will mean that small, marginal farming areas in other parts of the State can be opened up using its technique of 'micro-irrigation'.

There are several other parties that have an interest in developments in Trickle Creek, and Irrigation Inc's application to the State Water Commission.

State Water Commission (SWC)
The SWC has a mandate to develop the State's water resources, and has funds to support development. It is responsible for the administration of development and irrigation permits, and enforcing relevant State legislation.

Meatloaf Creek Irrigators
This group of irrigators and farmers are deeply concerned at developments in neighbouring Trickle Creek. They have had a regulated water supply scheme for the last fifteen years which has given them the advantage in irrigating because they have a more reliable supply – especially in dry times. Their competitive edge in the local market would vanish if the Trickle Creek scheme goes ahead. Their concern centres on the loss in property values and the markets on which they presently have a monopoly. If, however, the district could be restructured to produce a larger yield of agricultural produce, it may be feasible to expand into new markets – for instance, South-east Asia.

Didgereedoo Shire
The Shire Council is responsible for the management of the local area. The Council is keen to see development in the region. It has had Trickle Creek earmarked for residential/rural development, but could be persuaded that irrigation of the area presents a realistic and equally profitable alternative. The Shire will be represented by the Shire Clerk who has spearheaded a drive for development in the Shire.

The State Environmental Council (SEC)
This umbrella organisation represents all the environmental groups in the State. Trickle Creek does not have a formal environmental and conservation body, and so the SEC has undertaken the defence of the region's environment at the request of several prominent local residents. The main threat it sees is the damage to the region's natural ecosystems, particularly in the State Forest, as well as the destruction of the area's natural beauty.

The Water Advisory Committee (WAC)
The WAC is a group of local irrigators and water users that advises the State Water Authority on water use, restrictions and licence applications. It polices

the Authority's regulations, as well as its own restrictions; mostly armed with the weapon of peer-pressure. All issues that affect the water in the creek are discussed by the WAC and its members. Although the irrigators are pleased with the project in general because it provides a measure of security against drought, they are concerned with water quality and the price they will have to pay for their water. WAC members are evenly divided however between those who wish to preserve the natural beauty of the area and its relaxed lifestyle, and those who seek to develop the region and the markets for its produce. The WAC will be represented by the WAC Chairman – coincidentally the largest landholder in Trickle Creek.

The Issues

Irrigation Inc has already had several discussions with each of the main players in this project. As a result of these discussions, Irrigation Inc has identified five issues of concern to some or all of the parties. A general description of each is given below, with more detailed information on each party's views included in their briefing notes.

Damage to the 'slip country' and ecology

The 'slip country' and the parts of the State Forest reserve in the Trickle Creek headwaters will be cleared of all vegetation to make way for the two dams. The earth will then be formed into the dam walls. Irrigation Inc expects that grasses will grow back over the area within two years.

Local resident groups and the environmentalists have demanded much tighter environmental controls on this aspect of the development, however. They want to retain the natural beauty of the region and minimise erosion and lowering of water quality in the building stages.

Three options are open to solve this controversy:

1   **No regeneration.** This is the option that Irrigation Inc has submitted. In this scenario, no replantings of grass or native species would take place and vegetation would be left to regrow naturally. Some of the soil on the dam sites may wash, but Irrigation Inc argues that this will cease within two years of the establishment of the dam.

2   **Partial restoration.** Grass and some tree seedlings would be planted.

3   **Total regeneration.** Stands of native forest and vegetation would not be removed, and local tree species, grasses and understorey would be re-established in the catchment. Grasses and shrubbery native to the region would also be re-established. Of the three options, this is the most expensive financially. It may also be necessary to alter the dam design to cater for the altered catchment and runoff patterns.

Water quality

In the initial stages when the dams are being dug and established, and while vegetation regrows, Irrigation Inc admits that there will be some soil solids suspended in the water, but does not believe that it will affect water use significantly. Local irrigators and residents disagree. The amount of soil in the water could render it unfit for domestic and stock use for up to two years. (This is an upper limit. A more conservative estimate would be 12 months.) There are several options available:

1 **No controls** on water discharged into the creek.
2 **Some anti-erosion furrows** ploughed in the catchment to reduce erosion.
3 **Contour furrows and a settling pond** at each of the dam sites to filter soil solids before they enter the creek. The Shire Engineer believes that this will be sufficient to render water acceptable for irrigation and stock use.
4 **Contour furrows, settling ponds and filtration units** for those farmers who use the creek for domestic water.

Payment for supply

Irrigation Inc wants to charge the full amount of water costs to supply irrigators, together with a contribution (spread over thirty years) for the construction of the dams. Trickle Creek farmers – used to a free supply – are outraged at the price they are expected to pay. Irrigation Inc points out that with their increased profits due to a secure water supply, the farmers will easily be able to afford their charges. Without some form of payback it will be impossible for it to operate. Four options appear open;

1 **Full payment** of costs and construction charges
2 **75 per cent payment** of costs and construction charges
3 **50 per cent payment** of costs and construction charges
4 **25 per cent payment** of costs and construction charges.

Compensation for Meatloaf Creek irrigators

Irrigators on Meatloaf Creek are concerned about the developments in the neighbouring Trickle Creek catchment. Meatloaf Creek has long had the advantage in that it does have regulated water supply, albeit an inefficient and costly one. Some supply from this project is pumped into the lower reaches of Trickle Creek. This has meant that Meatloaf Creek can produce products more consistently and of higher quality. This advantage could be lost if Irrigation Inc's proposal goes ahead. Farmers, who have paid higher prices for their land and water, feel that they are due some form of compensation, at least for the first five years of operation while they restructure their local economy. They believe that they could lose up to $3 million a year in lost markets and property devaluation. Irrigation Inc points out that with the water that they will no longer be pumping into the lower parts of Trickle Creek, they will be able to grow crops more intensively, and have a greater safety margin in time of drought. Four options for compensation are open:

1   $3 million per year (100 per cent compensation) paid for the next five years

2   $2 million per year (67 per cent compensation) paid for the next five years

3   $1 million per year (33 per cent compensation) paid over the next five years

4   No compensation.

Loan for the building
The charter of the State Water Commission includes a mandate to promote the development of the State's water resources for agricultural and irrigation purposes. It can provide a government loan to support the development of the Trickle Creek project.

Irrigation Inc estimates the cost of the establishment of the Trickle Creek project to be $10 million, and has applied to the Commission for a loan, to be repaid over ten years, of $6 million. The Commission has indicated that there are several conditions that must be met before it can approve any grant. There are four options open to the Commission:

1   a $6 million loan, repaid over 10 years
2   a $4 million loan, repaid over 10 years
3   a $2 million loan, repaid over 10 years
4   No loan.

The Negotiation Process

Irrigation Inc has already applied to the State Water Commission for a permit to allow it to proceed with the Trickle Creek development. Its application proposed:

- no revegetation of the site
- no controls on soil into the creek
- full payment by users for water
- no compensation of Meatloaf Creek irrigators
- a $6 million loan, repaid over ten years.

Irrigation Inc is free to submit changes to its proposal, but is clearly keen to see the application approved in its present form.

In a bid to gain support for its proposal, Irrigation Inc has invited all the key parties to a meeting in the State Capital. The aim of this meeting is to find agreement between all parties and ensure unanimous support (although Irrigation Inc only needs the support of any four of the other parties).

All parties have agreed to come to the meeting, and have just met in the conference room. Each party has seen a copy of Irrigation Inc's proposal as well as the options that are open to them. The representative of Irrigation Inc will chair the meeting.

Discussions and negotiations may move in any direction, but Irrigation Inc will be looking for a proposal that will give it the support it needs to get approval from the Commission. Three formal voting rounds are scheduled during the meeting to determine the level of support for the proposal. The first will take place 15 minutes after the meeting begins, the second after 40 minutes of discussion and the final after 75 minutes. Additional votes may be taken at any stage, but at least three votes must be taken (unless the project receives enough votes for SWC approval).

The Commission representative will administer the three votes. If Irrigation Inc cannot decide on a revised proposal at the time of a formal vote, the participants must vote on the original proposal.

Voting is done by a show of hands. Once a proposal is passed (ie receives the support of at least five parties including Irrigation Inc) the votes are binding and parties cannot withdraw from their positions. The parties are free to explore 'improvements' in the agreement, which benefit the supporting parties or entice non-supporting parties to vote for agreement. If the proposed improvements are not unanimously agreed to by the parties to the original agreement, then the original agreement stands.

Negotiations must cease at the end of the meeting. If no agreement has been reached, then the State Water Commission will reject Irrigation Inc's application.

Irrigation Inc

To: The Negotiator
From: The Board of Directors,
            Irrigation Inc
Status: CONFIDENTIAL

This is clearly an important project to us. It will be the first irrigation development this company undertakes and we are keen to see our proposals accepted unchanged. As you are aware, several detailed studies into suitable irrigation sites have been undertaken, as well as intensive research into the 'micro-irrigation' techniques. Now it is time to make this project pay off.

The project has great profit potential, both for us and the Trickle Creek area. In the longer term, if the project is successful, as we expect it to be, there will be the opportunity for further developments in the rest of the State.

The benefits of the project seem so obvious to us that we are quite surprised at the hesitancy of the State Water Commission. However, the politics of the situation demand that we negotiate with organisations and individuals that have no business in our business. We feel that the application we made to the SWC six weeks ago represents a sound balance between environmental and economic considerations. We obviously want this package to be accepted and expect you to do your best to achieve this.

Scoring

To help you plan your negotiating strategy, the Board has developed a 20 point scoring scheme to help you determine those issues which are of greater or lesser significance to us. A score of 20 represents the most attractive solution to us, a score of 1, the least attractive. Compromising on one or more of the issues will earn a score somewhere between 1 and 20.

The idea of a point system may seem artificial and inflexible. Although the system obviously has limitations, it does allow us to combine several interests – costs, funding, environment and water quality – so that we can determine the overall merits of a package. The score will also allow us to determine the relative merits of one alternative over another. The higher the score, the more attractive the proposition. Note that if the project is to go ahead in any form, you must negotiate a package which will score at least eleven out of twenty. Any less than this and we would be better off looking at other sites, probably interstate.

Your task is to earn as many points as possible in this negotiation. This is not being greedy; it represents the best scenario in which we can develop the Trickle Creek site. We will support an agreement that scores at least 11 points – but that is the absolute minimum we can accept. We expect that you will be able to do much better than this.

Keep this information CONFIDENTIAL. Do not show the scoring system to anybody as the information may seriously jeopardise your negotiation.

The Issues

Payment
This is the most important issue to us. Without payment by the users of the system of what it will cost us to run this project, we will not be able to operate

as a private company. We can expect some support from the SWC, which is adopting a similar 'user-pays' principle. This makes them a natural ally in negotiations on this point. However, this does mean that the Commission cannot be relied on to subsidise the project.

We have calculated in real terms what it will cost us to build the dams and channels, and to supply water to the irrigators. To this we have added a profit margin to raise capital for further developments. Naturally, we would like to see our payments fee accepted as it is. Payment of 75 per cent of what we are asking will leave us with only a very narrow profit margin, and severely underfunded for future developments. Payment of only 50 per cent could only be supported in the long-term if future developments are profitable: they will have to subsidise this venture.(We can use Trickle Creek as an example of our irrigation technology, and as the basis for future, more profitable developments.) This is a very risky business however, and an option which you will have to avoid as far as possible. The last option of farmers only paying 25 per cent of the fee we intend to charge is not viable.

We can expect stiff opposition from the farmers. They have been used to free water from Trickle Creek for years now. It is worth stressing that by using the dams and micro-irrigation technology, they will be vastly more secure from drought, and will have a steady, reliable source of water with which to irrigate their crops and pasture.

Farmers will also try to compare our rates to those of Meatloaf Creek irrigation scheme. Meatloaf Creek is still heavily subsidised by the SWC, although there are moves to make the water fees charged there representative of the cost of providing the service. Meatloaf Creek has never been a paying proposition – let alone a profitable one – and with the irrigation techniques being used there, it never will be. Micro-irrigation technology could make an improvement, but we must show that the technique can work for Trickle Creek.

Irrigation Inc is not a charity and needs to make a working profit to support its activities, both now and in the future.

The Board has therefore drawn up the following points for this issue:

- Full payment of costs and construction charges: *8 points*
- 75 per cent payment of costs and construction charges: *6 points*
- 50 per cent payment of costs and construction charges: *2 points*
- 25 per cent payment of costs and construction charges: *0 points*

The points reflect the undesirability for the company for the last two options (50 per cent and 25 per cent payment)

Loan

Irrigation Inc is only a new company, and despite the support of several leading financial houses, commitment from the government to this project is important to us. Although we can raise the capital for this project, $10 million is a large sum if it is not backed by those responsible for water management in this State. This is especially important if we have to cut back on the amount we are paid for the service we provide, and have to rely on future projects to support us. We have asked the SWC for a loan of $6 million dollars, to be repaid over ten years. Irrigation Inc would provide the other 40 per cent of the initial funds needed in this development.

The SWC was initially quite positive about being able to support us on this basis, but since the Commission was roasted in Parliament over its funding procedures, it has been much more hesitant.

However, the Commission does have a mandate from State Parliament to develop the States' water resources, in particular for agriculture and irrigation, so we do feel confident of some support. Also, we would not like to create the precedent of a private company entirely underwriting what is, at the moment, a Commission responsibility.

On this basis we have arrived at the following points scheme for the loan we are seeking:

- $6 million loan, repaid over 10 years: *4 points*
- $4 million loan, repaid over 10 years: *3 points*
- $2 million loan, repaid over 10 years: *2 points*
- no loan: *0 points*

Compensation
Clearly we are opposed to paying compensation to the irrigators in Meatloaf Creek. We do not want to create the very dangerous precedent of compensating people every time someone loses business to us. In the same vein, why should we pay for our 'competitors' to restructure their business? It's clear where both of these scenarios could lead us.

The Meatloaf Creek irrigators have made extravagant claims on the amount that they are likely to lose in terms of lowering property prices and the loss of markets. The estimate that they have submitted to us claims for a loss of $3 million a year – for all of which they want compensation – for the next five years. This is totally unreasonable, and even if Trickle Creek were totally profitable, is more than the amount we are likely to make from it. Even $1 million compensation a year is a large slice of our operating capital.

Accordingly, we feel that we cannot provide any more than $1 million per year in compensation. We hope that you will be able to convince the other parties that this claim is not in the long-term interests of Trickle Creek or the company – and by extension, to the ongoing development of micro-irrigation technology in Australia (we hold the rights for MIT in Australia).

- $3 million per year compensation: *0 points*
- $2 million per year compensation: *1 points*
- $1 million per year compensation: *2 points*
- no compensation: *3 points*

Revegetation of the native ecology

The environmental lobby drag this issue up every time a development occurs. They claim that the area would be 'irreparably damaged', that the 'ongoing viability of the region would be seriously compromised', that the clearing of State Forest is 'illegal and immoral', and a host of other complaints. Most can be readily dismissed. Our advanced computer modelling shows that the site will stabilise within twelve months, six if there is a good growing season. Furthermore, revegetating is neither necessary or economic.

We do not want native forest regrown in the catchment, as it will affect water quality and runoff rates. This is the reason we have to clear it in the first place. This company is not run by ecological vandals, but is concerned with the proper, effective and economical management of our natural resources.

Environmental arguments tend to be emotional arguments however, and have a habit of making headlines. We do not need a label which we neither want nor deserve. We do concede however that it may be necessary to make at least some concessions in terms of replanting – especially if it boosts our image of an ecologically sensitive company.

The following points have been assigned to this issue:

- No revegetation: *3 points*
- Partial revegetation: *2 points*
- Total revegetation: *1 point*

Water quality

When the plans for the development were first released, farmers were instantly up-in-arms about what they felt were excessive levels of clay and soil particles in their water. What they based this claim on is not at all clear. It is not impossible that the WAC hired a consultant to 'massage' the figures.

It is true that when the dam sites are first dug, they will be prone to a small amount of erosion, but not enough to make irrigation impossible. After six to twelve months, the area will have grown over, and erosion and soil washing into the creek will cease. All this was contained our submission to the Commission, backed up by various scenarios generated by computer models. As far as we are aware, all the farms in the area get their domestic water from rain collected off their roofs, and water quality should be perfectly adequate for stock and irrigation.

It is not a particularly difficult task to build contour banks to stop soil washing, and settling ponds to collect soil that does wash; these can be built at the same time as the dam walls. However, the ponds will need to be removed at a later date – another expense. What we are keen to avoid is providing each farmer with a filtration unit. We would have to hire these on a long-term basis, and pay for their running costs – this could rapidly become a very expensive exercise. On this basis we have assigned the following points to this issue:

- No control: *2 points*
- Contour drains: *2 points*
- Contour drains and settling pond: *1 point*
- Contour drains, settling ponds and filtration units: *0 point*

CONFIDENTIAL SCORESHEET FOR IRRIGATION INC

Issue	Points 1st vote	2nd vote	3rd vote
1 Revegetation			
a No revegetation	3		
b Partial revegetation	2		
c Total revegetation 1			
2 Water quality			
a No controls	2		
b Contour furrows	2		
c Furrows and ponds	1		
d Furrows, ponds and filtration	0		
3 Payment			
a 100 % payment	8		
b 75 % payment	6		
c 50 % payment	2		
d 25 % payment	0		
4 Compensation			
a $3 million	0		
b $2 million	1		
c $1 million	2		
d no compensation	3		

5 Loan

a	$6 million loan	4
b	$4 million loan	3
c	$2 million loan	2
d	no loan	0

TOTAL

Minimum needed for agreement is 10 points out of 20

Summary of points for Irrigation Inc

Your name:

Did Irrigation Inc get the agreement it needed?
YES   NO

If 'yes', which parties agreed to the proposal?

State Water Commission   Didgereedoo Shire
State Environmental Council   WAC
Meatloaf Creek Irrigators

What was the agreement, which options were agreed to and how many points were generated?

Issue	Outcome	Points
1 Revegetation		
2 Water quality		
3 Payment		
4 Compensation		
5 Loan		

Total points

Irrigation Inc

> *To:*   The Negotiator
> *From:*   The Board of Directors,
>   State Environment Council
> *Status:*   CONFIDENTIAL
> Instructions to the Negotiator from the
> State Environment Council.

We are particularly concerned about Irrigation Inc's proposal to construct an irrigation scheme in the Trickle Creek area. The damage done in the construction of the dams and the rerouting of the creek will be considerable. That the Water Commission is prepared to sanction the use of State forest for the dam site only indicates this Government's priorities when it comes to development and environment.

Of the five issues to be discussed today, only two are of interest to the Environment Council: the level of environmental disturbance and regeneration, and the quality of the water flowing through and out of the creek. We have no strong feelings on the other three issues, although you will need to treat these issues carefully as your support – or otherwise – for them could win or lose allies. It is probably in your best interests not to disclose our position on these issues.

Scoring

To help you plan your negotiating strategy, the Board has developed a 20 point scoring scheme to help you determine those issues which are of greater or lesser significance to us. A score of 20 represents the most attractive solution to us, a score of 0, the least attractive. Compromising on one or more of the issues will earn a score somewhere between 0 and 20.

The idea of a point system may seem artificial and inflexible. Although the system obviously has limitations, it does allow us to combine several interests – the effects of development, the level of regeneration, and the quality of water – so that we can determine the overall merits of a package. The score will also allow us to determine the relative merits of one alternative over another. The higher the score, the more attractive the proposition. Note that we cannot afford to support any project that earns less than 11 points out of 20. If the project turns out to be an environmental disaster, we don't want the Environment Council's name associated with it. If Irrigation Inc does not proceed with the project, then the area will be left in its present unspoilt condition. If we do lend our support, we may be able to negotiate some environmental protection.

Your task is to earn as many points as possible in this negotiation. This is not being greedy; it represents the best scenario for the sensible management of Trickle Creek. We cannot support a scenario that earns less than 11 points and we expect that you will be able to do much better than this. You do not necessarily have to support a proposal that earns more than eleven – but the Environment Council is, however, keen to avoid the label of 'anti-development'. It is probably best not to block a proposal that you believe will be acceptable to the environment.

Keep this information CONFIDENTIAL. Do not show the scoring system to anybody as the information may seriously jeopardise your negotiation.

The Issues

Revegetation

The proposal Irrigation Inc has submitted to the State Water Commission indicates that the dam site will be left to regenerate 'naturally'. This is clearly another way of saying that Irrigation Inc has no intention of restoring the area. This is totally unacceptable. If the site were to suffer a major storm or flooding while it was still 'regenerating', large amounts of top soil would be lost, in which case, the area would no longer be able to support the local vegetation. We believe that it is possible to both build the dams and minimise the environmental disturbance, although this will require substantial revegetation being undertaken by Irrigation Inc.

Another issue is the stability of the dam sites. The region is called the 'slip country' because it is prone to landslides, especially in wet weather. The area is obviously going to be even more unstable if the surface vegetation – especially the tree cover – is removed.

The State Water Commission has come under fire in State Parliament for its handling of environmental issues, and now is much more concerned with this element of development. This does make its attitude of allowing State Forest to be cleared puzzling. Allowing the diversion of Trickle Creek is also unusual – both in ecological and engineering terms. We have allocated the following points to this issue:

- No revegetation: *0 points*
- Partial revegetation: *7 points*
- Total revegetation: *12 points*

Water quality

If large amounts of soil and sediment are washed into the river system, it has the potential to affect a much larger region than Trickle Creek. If the water becomes muddy, conceivably the entire ecological downstream river system could change (including the upper Sally River into which Trickle Creek flows).

Certainly the local ecology could change – and once the damage has been done, it will be very difficult to put right. Proper water quality control measures need to be in place from day one if this project is to go ahead.

We believe that we can rely on the farmers for support on this issue – although for quite different reasons (they will be looking for water for their crops and stock).

- No control: *0 points*
- Contour drains: *2 points*
- Contour drains and settling pond: *7 points*
- Contour drains, settling ponds and filtration units: *8 points*

Other issues
We have no interest in the other three issues – the subsidy for the development, the payments that farmers will make for water or the compensation that the Meatloaf Creek farmers are seeking. Accordingly, we have allocated no points to these issues. Treat these three issues as seems appropriate to you, but be careful not to antagonise any potential allies.

CONFIDENTIAL SCORESHEET FOR THE STATE ENVIRONMENT COUNCIL

Issue	*Points 1st vote*	*2nd vote*	*3rd vote*
1 Revegetation			
a No revegetation	0		
b Partial revegetation	7		
c Total revegetation	12		
2 Water Quality			
a No controls	0		
b Contour furrows	2		
c Furrows and ponds	7		
d Furrows, ponds and filtration	8		
3 Payment			
a 100 % payment	0		
b 75 % payment	0		
c 50 % payment	0		
d 25 % payment	0		

4 Compensation

a. $3 million	0
b. $2 million	0
c. $1 million	0
d. no compensation	0

5 Loan

a $6 million loan	0
b $4 million loan	0
c $2 million loan	0
d no loan	0

TOTAL

Minimum needed for agreement is 11 points out of 20

Summary of points for the State Environment Council

Your name:

Did Irrigation Inc get the agreement it needed?
YES   NO

If 'yes', which parties agreed to the proposal?

State Water Commission   Didgeredoo Shire
State Environmental Council   WAC
Meatloaf Creek Irrigators

What was the agreement, which options were agreed to and how many points were generated?

Issue	Outcome	Points
1 Revegetation		
2 Water quality		
3 Payment		
4 Compensation		
5 Loan		

Total points

Irrigation Inc

Instructions from the Didgeredoo Shire Council

As you are aware, several years ago the Shire embarked on a project of development, aimed at improving the profitability, value and facilities of the area. Also, we want to see the area developed for the benefit of all – not just a minority. So far, the project has been slow to start. In the present financial climate, funds and investors are in scarce supply. The council has re-zoned some areas for residential development, improved roads in the area, and extended the Shire's main shopping and supply centre. However, things have picked up recently: there has been the Irrigation Inc proposal for Trickle Creek, and some interest in a rural housing development in the same catchment. This leaves us with something of a dilemma.

The Trickle Creek project involves some difficult decisions for Didgeredoo Shire Council. We had originally zoned the area for a very low-density, semi-rural housing estate. There has been some interest in this move from several large property developers, but as yet, no firm commitment has been made by any of them. A rural housing development would generate additional revenue for the Shire (through rates), as well as bringing investment into the region.

Not surprisingly, the concept of a rural housing development is not popular amongst the farming community it will displace. They have strong – and legitimate – claims on the area. However, Trickle Creek is at best only marginally profitable. For several years, prices for vegetables and fruit have been less than it costs to produce them.

The Trickle Creek farmers have had an offer from a supplier to the Asian fruit and vegetable market. They will have to guarantee both supply and quantity. Supply can only be guaranteed if the Irrigation Inc proposal is approved; quantity if they combine their produce with what is grown in Meatloaf Creek.

As far as the Council is concerned, we are pleased that this is one development that will require little input from us. It will not cost us anything beyond normal planning and inspection costs. The by-laws do not have to be altered to allow this sort of development, nor do the present zoning restrictions. It is also a project that will be popular with irrigators in Trickle Creek because it assures them water security – despite the fact that it is going to cost them. It will also see land values in Trickle Creek improve.

We do have some reservations however. The 'slip country' where the dams are to be built is not very stable – there are several landslides every year. We would need guarantees that the whole project will be safe (we don't want even the risk of a dam collapsing). The area is also naturally beautiful – hence the attraction of turning the area into a rural housing estate. Any development must not spoil the area. At present, Irrigation Inc's proposal does not include any provision for regrowing the native forest area, or even revegetating the dam sites. This is clearly not acceptable. The Shire Engineer believes that it should be possible to build the dams without disturbing the environment – especially if the area is revegetated at the same time as construction.

A related issue is the amount of soil washed into the creek, and into the upper Sally River (which flows through the rest of the Shire). The Trickle Creek farmers are claiming that they would be unable to use that water for irrigation for up to twelve months. We think that their claim is wildly exaggerated. What is not wildly exaggerated, however, is the assertion that the water will not be acceptable for stock or domestic use – both in Trickle Creek or in the Upper Sally River near the development – for several months. We obviously cannot allow farmers to be left without livelihood or stock without adequate water for several months, especially those outside Trickle Creek and who will not benefit from the project. The Shire Engineer has advised us that contour banks dug on the dam site and a settling pond to collect soil before it enters the creek should be sufficient to ensure the water quality for irrigation and livestock.

Meatloaf Creek also poses us with something of a dilemma. The farmers there have lobbied the council quite forcefully to disallow the Trickle Creek project. If we won't do that, they want us to press for compensation for the loss of property values and markets. This is a more serious problem. We do not want to support a project that will have the effect of decreasing the overall value of the region. The compensation that the farmers are claiming ($3 million per year for the next five years) is a gross overestimate of the value of their land and market, and could have the effect of scaring off Irrigation Inc. The Council feels that a more moderate line should be adopted. Compensation will allow them to retool, and hopefully combine with Trickle Creek to establish a niche in the Asian market.

Scoring

To help you plan your negotiating strategy, the council has developed a 20 point scoring scheme to help you determine those issues which are of greater or lesser significance to us. A score of 20 represents the most attractive solution to us, a score of 1, the least attractive. Compromising on one or more of the issues will earn a score somewhere between 1 and 20.

The idea of a point system may seem artificial and inflexible. Although the system obviously has limitations, it does allow us to combine several interests – the effects of development, the level of regeneration, and the quality of water – so that we can determine the overall merits of a package. The score will also allow the Shire Council to determine the relative merits of one alternative over another. The higher the score, the more attractive the proposition.

The scoring system will also help you balance which of the developments we should proceed with. If the project does not go ahead – fine. We will continue on our original course of attracting developers to the region to create the rural housing project. If it does go ahead, we have a duty to represent Trickle Creek farmers, the concerns of irrigators in Meatloaf Creek, and those who may be affected by silting in the Upper Sally River.

Our bottom line is that we are aiming to attract developments that are of greatest benefit to the Shire. Irrigation Inc is the first major development that has eventuated in the Shire since we began our promotion of the area for development. Given the present economic climate, it may the be last for a while, especially if it goes away from the meeting without an agreement.

Your task is to earn as many points as possible in this negotiation. This is not being greedy; it represents the best scenario in which we can develop the Trickle Creek site. We will support an agreement that scores at least eight points out of twenty – but that is the absolute minimum we can accept. We expect that you will be able to do much better than this.

The Issues

Revegetation
The beauty of the area needs to be preserved, along with the stability of the dam sites. Also, the construction sites must not be allowed to erode and wash soil into Trickle Creek or the Upper Sally River.

- No revegetation: *0 points*
- Partial revegetation: *3 points*
- Total revegetation: *4 points*

Water quality
This is a major source of concern for us regarding this project, as poor quality water flowing out of Trickle Creek could affect the rest of the Shire. Apart from affecting residents and users, it could cost the council a small fortune in lost crops and damage to the waterway. The Shire Engineer has said that the level of suspended solids in the water would be unacceptable, and has relayed this opinion to the Trickle Creek WAC.

- No control: *0 points*
- Contour drains: *4 points*
- Contour drains and settling pond: *5 points*
- Contour drains, settling ponds and filtration units: *6 points*

Payment

The irrigators in Trickle Creek are upset at the cost they will be charged for water under the new scheme. The view of the council is that there should be a parity of charges between Trickle Creek and Meatloaf Creek. If charges are higher, then it will decrease the profitability of the Trickle Creek area, if they are less, we may lose Irrigation Inc.

- Full payment of costs and construction charges: *1 point*
- 75 per cent payment of costs and construction charges: *2 points*
- 50 per cent payment of costs and construction charges: *3 points*
- 25 per cent payment of costs and construction charges: *2 points*

Compensation

The farmers and irrigators have lobbied the council persistently for compensation. The view of the council is that, although they may be losing some value to property and markets, it is nowhere near what they are claiming. In any case, their claim may scare off Irrigation Inc. By playing for a middle line, the council hopes to make both areas profitable – Trickle Creek through irrigation, Meatloaf Creek by restructuring. Also, in the long term, an injection of funds will help to establish new markets for the region's produce. An increase in profitability in both areas will flow out into the rest of the Shire.

- $3 million per year compensation: *1 point*
- $2 million per year compensation: *2 points*
- $1 million per year compensation: *3 points*
- no compensation: *0 points*

Loan

This is not an issue of direct relevance to us. Our main interest is seeing that the project does not collapse on financial grounds. Support from the State Water Authority would ensure this (the government is unlikely to walk away from several million dollars worth of investment). Government investment will also ensure that Didgeredoo Shire is not left carrying the can if the project does fail.

- $6 million loan, repaid over 10 years: *4 points*
- $4 million loan, repaid over 10 years: *3 points*
- $2 million loan, repaid over 10 years: *2 points*
- no loan: *0 points*

CONFIDENTIAL SCORESHEET FOR DIDGEREDOO SHIRE COUNCIL

Issue	Points 1st vote	2nd vote	3rd vote
1 Revegetation			
a No revegetation	0		
b Partial revegetation	3		
c Total revegetation	4		
2 Water Quality			
a No controls	0		
b Contour furrows	4		
c Furrows and ponds	5		
d Furrows, ponds and filtration	6		
3 Payment			
a 100 % payment	1		
b 75 % payment	2		
c 50 % payment	3		
d 25 % payment	2		
4 Compensation			
a $3 million	1		
b $2 million	2		
c $1 million	3		
d no compensation	0		
5 Loan			
a $6 million loan	4		
b $4 million loan	3		
c $2 million loan	2		
d no loan	0		

TOTAL

Minimum needed for agreement is 8 points out of 20

Summary of points for the Didgeredoo Shire Council

Your name:

Did Irrigation Inc get the agreement it needed?
YES   NO

If 'yes', which parties agreed to the proposal?

State Water Commission   Didgeredoo Shire
State Environmental Council   WAC
Meatloaf Creek Irrigators

What was the agreement, which options were agreed to and how many points were generated?

Issue	Outcome	Points
1 Revegetation		
2 Water quality		
3 Payment		
4 Compensation		
5 Loan		

Total points

Irrigation Inc

Instructions to the Meatloaf Creek Negotiator

The Trickle Creek Irrigation Scheme proposed by Irrigation Inc is of great concern to us, the irrigators and farmers of Meatloaf Creek. We have hired you to represent our interests, and to do all you can to stop this project going ahead.

An irrigation scheme was established on Meatloaf Creek about fifteen years ago. This has provided us with a fair degree of security against drought and dry spells. It has meant that we can produce crops similar to those of Trickle Creek, but more consistently. A stable water supply has also meant generally better quality produce. Together, these have ensured us a much better market for our crops locally. Irrigation Inc's proposal would severely affect our profitability by introducing a competitor.

We are also concerned with the effects on property prices in Meatloaf Creek if the project does go ahead. Most of the farmers who bought into the scheme paid higher prices than they would have for an equivalent block in Trickle Creek (because of the provision of a regulated water supply). If the Trickle Creek project is approved, the security that these farmers bought will be less valuable in this Shire.

We had an auditor assess what we were likely to lose if the project went ahead. She told us that it was very difficult to assess as it depended a lot on the success of the Trickle Creek project, as well as on things like property markets and the sale of produce. Looking at everything, she told us that, at worst, we could expect to lose approximately $1.9 million per year as a region. We have set our compensation claim higher than this, as we expect to have it cut back in any case. We will use whatever we can get from Irrigation Inc to restructure our local markets and explore new ones, and to improve our irrigation systems.

Three weeks ago, the irrigators of Meatloaf Creek met to work out what we saw as the issues affecting us. We believe that the project should not go ahead, but we suspect we are in the minority and that it will be given approval. If this is the case, then our main priority is compensation for our loss of income. Several farmers did bring up other issues at this meeting, and we agreed to include them in our list of concerns – even if we can't stop the project, we might be able to ensure that it does the least possible harm.

To help you plan your negotiating strategy, we developed a 20 point scoring scheme to help you determine those issues which are of greater or lesser significance to us. A score of 20 represents the most attractive solution to us, a score of 0, the least attractive. Compromising on one or more of the issues will earn a score somewhere between 0 and 20.

The idea of a point system may seem artificial and inflexible. Although the system obviously has limitations, it does allow us to combine several interests – the level of compensation, the effects of the development locally, and the level the SWC would subsidise the project – so that you can determine the overall merits of a package. The score will also allow you to determine the relative merits of one alternative over another. The higher the score, the more attractive the proposition.

Our feeling is that the project should not go ahead at all, and if you can manage to achieve this goal, we will pay a bonus on your fee (this is equivalent to 30 points).

Your task is to earn as many points as possible in this negotiation. This is not being greedy; it represents the best set of options for us and our future. We feel that, given the apparent level of support for the project from the Shire Council and the SWC, our best strategy may be to aim to get as much compensation as possible. You may however be able to ensure that the project does not get off the ground at all. If you can get more than 8 out of 20, we feel it is better to support the proposal and get as much compensation as we can. We can't support a proposal that earns less than 8 points.

The Issues

Revegetation

Although the state of the dam sites and the water quality are not issues that affect us directly, we cannot believe that Irrigation Inc would be allowed to develop this scheme without some protection of the environment. The whole Shire is naturally beautiful and nothing should be allowed to destroy that, so the more protection and revegetation of the dam sites, the better. We believe that it is technically possible to build and operate the dams while minimising the damage – especially to the State Forest. We feel that the water flowing into the Upper Sally River should be of a quality comparable to the present.

- No revegetation: *0 points*
- Partial revegetation: *1 point*
- Total revegetation: *3 points*

Water quality

- No control: *0 points*
- Contour drains: *1 point*
- Contour drains and settling pond: *2 points*
- Contour drains, settling ponds and filtration units: *3 points*

Compensation

This is far the most important issue to us. In the event of you not being able to stop the project, we feel that we need to receive compensation. We have decided to use this money to restructure our local markets and improve our irrigation system. This will make us more competitive. We have lobbied Didgeredoo Shire Council for their support on this issue, and we believe we may have a sympathetic ear there – despite the fact that they have a pro-development bias. We believe that we have been able to convince them that there is no point in this development if the overall effect on the Shire is not in its everyone's best interests. On this basis we have drawn up the following points to represent the level of compensation we feel we are entitled.

- $3 million per year compensation: *7 points*
- $2 million per year compensation: *5 points*
- $1 million per year compensation: *3 points*
- no compensation: *0 points*

Payment

We are aware that the State Water Commission is pushing to have irrigators pay for what it *claims* is the full cost of water supply and repay the cost of the dams and channels that were built thirty years ago at today's prices. Our concern is that if Irrigation Inc charges the fees that it is proposing, it will give the SWC a good excuse to increase their prices. As far as we are concerned, the rates we are charged are already extravagant. If the Meatloaf Creek project has never been profitable (as the SWC claims) then it is because of SWC mismanagement, not through a lack of money or undercharging. On this basis, we are prepared to accept that if the Trickle Creek project must go ahead, then its users should be charged what we are charged now.

- Full payment of costs and construction charges: *0 points*
- 75 per cent payment of costs and construction charges: *0 points*
- 50 per cent payment of costs and construction charges: *1 point*
- 25 per cent payment of costs and construction charges: *4 points*

Loan

We are concerned about the amount of financial support that Irrigation Inc is seeking from the State Water Commission. The brief that Irrigation Inc supplied us with said that it was in a position to fund all of its own projects. Why does it need a loan? We feel that Irrigation Inc must be financially secure, because the region cannot afford for it to collapse and drag down Trickle Creek – especially if it has demolished us already.

If the SWC does provide Irrigation Inc with the funding it is asking for, it will effectively gain even more control over the use and cost of water in the area – something we are clearly opposed to.

- $6 million loan, repaid over 10 years: *0 points*
- $4 million loan, repaid over 10 years: *0 points*
- $2 million loan, repaid over 10 years: *1 point*
- no loan: *3 points*

CONFIDENTIAL SCORESHEET FOR MEATLOAF CREEK IRRIGATORS

Issue	Points 1st vote	2nd vote	3rd vote
1 Revegetation			
a No revegetation	0		
b Partial revegetation	1		
c Total revegetation	3		
2 Water Quality			
a No controls	0		
b Contour furrows	1		
c Furrows and ponds	2		
d Furrows, ponds and filtration	3		
3 Payment			
a 100 % payment	0		
b 75 % payment	0		
c 50 % payment	1		
d 25 % payment	4		
4 Compensation			
a $3 million	7		
b $2 million	5		
c $1 million	3		
d no compensation	0		
5 Loan			
a $6 million loan	0		
b $4 million loan	0		
c $2 million loan	1		
d no loan	3		

TOTAL

Minimum needed for agreement is 8 points out of 20
Bonus of 30 points for no agreement

Summary of points for the Meatloaf Creek Irrigators

Your name:

Did Irrigation Inc get the agreement it needed?
YES   NO

If 'yes', which parties agreed to the proposal?

State Water Commission   Didgeredoo Shire
State Environmental Council   WAC
Meatloaf Creek Irrigators

What was the agreement, which options were agreed to and how many points were generated?

Issue	Outcome	Points
1 Revegetation		
2 Water quality		
3 Payment		
4 Compensation		
5 Loan		

Total points

Irrigation Inc

The Chairman of the WAC

By and large, the WAC is pleased by Irrigation Inc's proposal. It will mean a much more secure source of water during dry periods and droughts. This in turn means that we will be able to produce larger crops, more consistently, and hopefully of better quality. However, there are several aspects of the proposal that we object to.

Irrigation Inc intends to charge what can only be described as outrageous fees for water use. This is clearly a case of blatant profiteering. Once the facility is installed, we will all have to pay for the water – there will not be the option of pumping water from the creek's natural flow for free as happens now.

The charges that are being suggested are four times what irrigators in Meatloaf Creek are being charged. Irrigation Inc claims that its 'micro-irrigation technology' will mean that water use is much more efficient. If we are going to be using less water, why are we paying more for it?

At the moment, the WAC effectively has control over water restrictions, enforcing pumping times and licence applications (although technically, we simply advise the SWC, it has never ignored our advice yet). These responsibilities will go out of our hands – almost certainly into those of Irrigation Inc. Up until now, it has been the State Water Commission's policy to listen to the concerns and advice of those that use the State's water resources. Clearly the Commission is now more interested in big business than local farmers.

If we can resolve these problems satisfactorily, however, then the future looks good. The project will ensure us a secure source of water – something we have wanted for years. We will not have to abandon our crops and sell off our prime stock when conditions become too dry. We have already had a tentative offer from a feeder for the Asian market, if we can guarantee supply and quantity. If we could combine our produce with what is produced in Meatloaf Creek, we could fill this order. Unfortunately, Meatloaf Creek is very antagonistic toward the project. They think they are going to lose markets and their property values are going to drop. They've presented Irrigation Inc with an absurd compensation claim. All this is going to do is drive the developers away and we'll be right back where we started.

The development will have another important side effect: derailing the Shire Council's plans for a housing development in Trickle Creek. The area is prime agricultural land being run by people who understand it. The last thing we want is to break up productive properties, and turn them into hobby farms and worse, so that a load of city-born freeloaders can enjoy the benefits of 'a relaxed country lifestyle'. It would completely destroy the community that we have here at the moment, not to mention the jobs and profits Trickle Creek generates.

At the last meeting of the Trickle Creek WAC three weeks ago, the irrigators met to work out what we saw as the issues affecting us. The meeting developed a 20 point scoring scheme to help determine those issues which are of greater or lesser significance to us. A score of 20 represents the most attractive solution; a score of 0, the least attractive. Compromising on one or more of the issues will earn a score somewhere between 0 and 20.

The idea of a point system may seem artificial and inflexible. Although the system obviously has limitations, it does allow us to combine several interests – the cost of water, the effect on water quality, the effect on the 'slip country' and so on – so that we can determine the overall merits of a package. The score will also allow us to determine the relative merits of one alternative over another. The higher the score, the more attractive the proposition. If this project is to have our support however, it must attract at least nine points out of twenty. Any less than nine and the disadvantages of the proposal will outweigh the advantages.

Your task is to earn as many points as possible in this negotiation. This is not being greedy; it represents the best set of options for us and our future. We want to make sure that if the project does go ahead, then it is on acceptable terms, and that we and the district are not going to suffer as a result.

The Issues

Revegetation

An issue related to that of water quality is what happens to the dam sites – are they to be revegetated, and if so, to what extent? Some farmers argued very strongly at the last WAC meeting that the area should be preserved, and that Irrigation Inc could build the irrigation scheme and preserve the beauty of the area. The ideal situation for us is to construct the dams without damaging the environment – something the Shire Engineer believes is possible. Everybody agreed that we should not allow erosion to set in the 'slip' country – especially since it is so unstable anyway. The meeting decided that the more ground cover we could get onto the sites, the better. The points we decided on were:

- No revegetation: *0 points*
- Partial revegetation: *2 points*
- Total revegetation: *4 points*

Water quality

Irrigation Inc's proposal that it submitted to the SWC indicates that while the dam sites are being dug and are 'stabilising' there will be a fair amount of soil washed into Trickle Creek. The Didgeredoo Shire Council Engineer has told us that this will make the water unusable – for irrigation, stock and domestic use (although only a couple of farmers use the creek for domestic water these days). They claim that this will stop after about twelve months. We cannot afford to stop growing crops or watering cattle for twelve months just because Irrigation Inc does not want to do something about erosion. At the moment, the water in Trickle Creek is totally uncontaminated. It should stay that way, and Irrigation Inc should be made to do whatever necessary to ensure that the quality is not affected during construction. There are a range of options: building contour drains to stop water washing soil into the creek, 'settling ponds' to collect soil that does wash, and providing filtration units to those farmers that need them. The WAC's view is that we should have all of these if possible. On this basis, the WAC decided on the following point system for this issue.

- No control: *0 points*
- Contour drains: *1 point*
- Contour drains and settling pond: *2 points*
- Contour drains, settling ponds and filtration units: *3 points*

Payment

This is the most important issue as far as we are concerned. We probably could not afford to run our farms profitably given the charges that Irrigation Inc wants to charge. We would be better off risking drought and an unsure water supply, than paying what they are asking at the moment. Although we can appreciate that Irrigation Inc has a business to run, it should not be at our expense. We feel that it is unfair that we should have to pay more than what the Meatloaf Creek irrigators are paying at the moment – if we have to pay, it should be on equal terms. The meeting decided on the following system of points for this issue:

- Full payment of costs and construction charges: *0 points*
- 75 per cent payment of costs and construction charges: *1 point*
- 50 per cent payment of costs and construction charges: *3 points*
- 25 per cent payment of costs and construction charges: *6 points*

Compensation

This was an issue that the WAC meeting was divided on. Some farmers felt that Meatloaf Creek did not deserve any compensation. They had never compensated us when their dam was built: why should they receive compensation now? Others felt that we could be in the same situation some day, and we might need all the allies we can find, and that some support may be useful.

Also if we are going to try to expand our markets with them, we cannot afford to antagonise the Meatloaf Creek irrigators. In any case, collaborating is going to mean that they are going to need to update their equipment – for which they are going to need financial support. We may be able to get Irrigation Inc to support us both this way. This is almost certainly not what they have in mind however.

By and large, the meeting felt that we should support the compensation claim, although not to the scale that Meatloaf Creek would like to see. We feel that $1 million is acceptable. A huge claim could scare off Irrigation Inc – something that should be avoided if possible.

- $3 million: *1 point*
- $2 million: *2 points*
- $1 million: *3 points*
- no compensation: *0 points*

Loan

By and large, the WAC is in favour of the proposal, and so we felt that we should support Irrigation Inc when it's not going to hurt us. So although the loan is not really an issue for us, we felt that we should try and back Irrigation Inc up on his one. Also, if the SWC is financially committed to this project, it is less likely to fail on financial grounds – and the more it is in for, the less likely it is to walk away from it.

Also if the SWC is involved in the project this way, it will probably give us a lever to preserve the role of the WAC. We feel that it is an important part of the operation of irrigation on the creek, and control of water use should not go out of the hands of the users.

The points for this issue are:

- $6 million loan, repaid over 10 years: *3 points*
- $4 million loan, repaid over 10 years: *2 points*
- $2 million loan, repaid over 10 years: *1 point*
- no loan: *0 points*

CONFIDENTIAL SCORESHEET FOR TRICKLE CREEK WAC

Issue	Points 1st vote	2nd vote	3rd vote
1 Revegetation			
a No revegetation	0		
b Partial revegetation	2		
c Total revegetation	4		
2 Water Quality			
a No controls	0		
b Contour furrows	1		
c Furrows and ponds	2		
d Furrows, ponds and filtration	3		
3 Payment			
a 100% payment	0		
b 75% payment	1		
c 50% payment	3		
d 25% payment	6		
4 Compensation			
a $3 million	1		
b $2 million	2		
c $1 million	3		
d no compensation	0		

5 Loan

a	$6 million loan	3
b	$4 million loan	2
c	$2 million loan	1
d	no loan	0

TOTAL

Minimum needed for agreement is 9 points out of 20

Summary of points for the Trickle Creek WAC

Your name:

Did Irrigation Inc get the agreement it needed?
YES   NO

If 'yes', which parties agreed to the proposal?

State Water Commission   Didgeredoo Shire
State Environmental Council   WAC
Meatloaf Creek Irrigators

What was the agreement, which options were agreed to and how many points were generated?

Issue	Outcome	Points
1 Revegetation		
2 Water quality		
3 Payment		
4 Compensation		
5 Loan		

Total points

Irrigation Inc

From the State Water Commissioner
Subject: Irrigation Inc Development Proposal for Trickle Creek

The Commission is very keen to support Irrigation Inc's proposal for several reasons. We believe that they will be able to construct facilities more economically than our structure allows us. They can more easily justify charging full fees (which would then allow us to follow suit in Meatloaf Creek, which we operate). And we would be keen to see the Micro-Irrigation (MI) technology developed in this State.

Until very recently, Irrigation Inc's proposal would have found a very receptive response from the Commission. It is much in line with developments elsewhere around the world: a move to privatise the development of water resources. It aims to charge the full cost of what it costs to provide the water: a move the Commission is continually hampered in. However, as you are aware, the Commission has come under considerable pressure recently for its involvement in developments of this nature. We are keen to see that this situation is not prolonged.

The two main concerns that Parliament and the Minister have had are the quality of 'environmental management' and the way the Commission has distributed its funds. The second point boils down to the view that Parliament feels we are overspending in what is clearly a poor financial period. The first issue is the more difficult one.

Parliament has begun to view 'environment' in much broader terms recently. In our case, 'environment' now includes issues such as the community that uses the water, the crops, pastures and stock that the water resource is used on, the catchments – as well as the traditional issues of water quality and dam sites. This is all well and good in theory, but how it is to be figured into planning in any practical sense is not at all clear to us.

It needs to be emphasised, however, that the Commission will remain responsible for the use of the water. If permission is given to Irrigation Inc for this project to go ahead, it will be on these terms. This is worth spelling out at the meeting you will be attending. The WAC will have a place within this structure: it has proved to be a valuable source of advice and an effective watchdog on water use.

To help you plan your negotiating strategy, the Board has developed a 20 point scoring scheme to help you determine those issues which are of greater or lesser significance to us. A score of 20 represents the most attractive solution to us, a score of 1, the least attractive. Compromising on one or more of the issues will earn a score somewhere between 1 and 20.

The idea of a point system may seem artificial and inflexible. Although the system obviously has limitations, it does allow us to combine several interests – costs, funding, environment and water quality – so that we can determine the overall merits of a package. The score will also allow us to determine the relative merits of one alternative over another. The higher the score, the more attractive the proposition. Note that if the project is to go ahead in any form, you must negotiate a package which will score of at least 11 out of 20. Any less than this represents unacceptable use to the water resource.

Your task is to earn as many points as possible in this negotiation. This is not being greedy; it represents the best scenario in which the Trickle Creek site may be developed. We will support an agreement that scores at least 12 points – but that is the absolute minimum we can accept. We expect that you will be able to do much better than this.

Keep this information CONFIDENTIAL. Do not show the scoring system to anybody as the information may seriously jeopardise your negotiation.

The Issues

Revegetation

The Commission not only has responsibility for water but also the management of water catchments. Our engineers have some concerns with the removal of vegetation cover – especially tree cover – from the 'slip country'. The area suffers several landslides each year, especially during wet periods, and removing vegetation will only decrease the stability of the site. Also, we cannot allow the site to become eroded as this will decrease the water quality in the long term. The preservation of the State Forest reserve is also an issue at present – we cannot support a proposal that does not contain some element of environmental protection so soon after parliamentary criticism on the same issue. Accordingly, the Commission has established the following points scheme:

- No revegetation: *0 points*
- Partial revegetation: *2 points*
- Total revegetation: *4 points*

Water quality

Clearly, any development concerning the State's water resources must not degrade water quality in the long term. Irrigation Inc's proposal indicates that there would be soil particles from the dam sites suspended in the water. There are various claims on how long this will last – Irrigation Inc says six months; irrigators say up to two years. In either case, soil washed into the river for this length of time could not only affect the long-term state of Trickle Creek, but also the Upper Sally River (into which Trickle Creek flows). This is not acceptable, and the Commission would require Irrigation Inc to carry out at least

276

some works to ensure water quality in the region. We also have a duty to protect the legitimate interests of those using the water resource – in this case the Trickle Creek community. Those relying on the creek for domestic water will need to be provided with filtration units. On this basis, the Commission has allocated the following points to this issue:

- No control: *0 points*
- Contour drains: *1 point*
- Contour drains and settling ponds: *3 points*
- Contour drains, settling ponds and filtration units: *4 points*

Payment

The Commission has been trying to introduce a 'user-pays' charging system for water consumers. This way, they will be paying what it costs to provide water and associated facilities. At present, irrigators and other rural water users are heavily subsidised by the State Government. Not surprisingly, the Commission has met with considerable resistance to this move in rural areas. By supporting a private development that intends to charge the full cost of after supply, the Commission will then have a precedent for introducing realistic pricing levels. In particular, we will have a local precedent for improving the viability of the Meatloaf Creek irrigation scheme. The Commission will increase the water rates of Meatloaf Creek to match the charges of Irrigation Inc. As this will undoubtedly be unpopular locally, it is probably best not to reveal this prematurely – but, if for instance, the Trickle Creek irrigators are demanding parity with Meatloaf Creek charges, then the increase in charges could be announced.

Irrigation Inc has indicated to us that without charging a realistic price for the service it will be providing, it will not be able to afford to develop this site. Consequently, the Commission will support Irrigation Inc in their present charging arrangement.

- Full payment of costs: *4 points*
- 75 per cent payment of costs: *3 points*
- 50 per cent payment of costs: *2 points*
- 25 per cent payment of costs: *0 points*

Compensation

The Commission's view of the Meatloaf irrigators compensation claim is that this would be a good way to upgrade the Meatloaf Creek scheme with minimal financial input from us. If necessary, Irrigation Inc can offset the cost of compensation against a loan from the Commission. In this way, two catchments are being developed, not just one. Furthermore, the level of payment that the Trickle Creek irrigators are asking for ($3M) is significantly less than the development costs of Trickle Creek ($10M), an additional saving. With the introduction of Irrigation Inc's 'micro-irrigation' technology, the Commission is positive that Meatloaf Creek can become a paying proposition – something it has never been. It would be worth pointing out to the Irrigation Inc representative that if it contributed to the development of Meatloaf Creek, it would be entitled to a share of the profits flowing from the sale of water (although not as much as Trickle Creek as the Commission would retain ownership of the present dams and channels). Thus the Commission willingly supports the Meatloaf Creek claim as a means to increase the viability of the entire region.

- $3 million per year compensation: *3 points*
- $2 million per year compensation: *2 points*
- $1 million per year compensation: *1 point*
-   no compensation: *0 points*

Loan

The Commission finds itself caught between opposing demands on this issue. Our role is to develop the State's water resources – a role that requires funds however development is carried out. On the other hand, the Commission has received a lot of criticism – both in Parliament and in the press – over our funding programme. The Minister has given express instructions that we are to review our expenditure. In short we are being asked to do the same job for less money.

Irrigation Inc represents an opportunity to achieve this objective. It has stated that it has the capital to support its own projects, and will not be reliant on government subsidy. In seeking a loan from us for this project, Irrigation Inc is apparently seeking underwriting for this project: presumably so that a future minister will not be tempted to withdraw government support during the development phase. Also, private development of irrigation schemes such as this have little precedent in Australia, and so Irrigation Inc is naturally looking for some guaranteed support in its establishment. We would like to encourage this type of development as they are not a financial burden on the Commission, so long as it as they do not degrade water resources.

Balancing these priorities, the Commission has decided that it can provide a loan of $2 million, to be repaid over ten years. We would be in a position to provide a larger loan on the condition that Irrigation Inc undertook work to improve water and catchment quality. The points that have been allocated to this issue are :

- $6 million repaid over ten years: *1 point*
- $4 million repaid over ten years: *3 points*
- $2 million repaid over ten years: *5 points*
-   no loan: *2 points*

CONFIDENTIAL SCORESHEET FOR THE STATE WATER COMMISSION

Issue	Points 1st vote	2nd vote	3rd vote
1 Revegetation			
a No revegetation	0		
b Partial revegetation	2		
c Total revegetation	4		
2 Water Quality			
a No controls	0		
b Contour furrows	1		
c Furrows and ponds	3		
d Furrows, ponds and filtration	4		
3 Payment			
a 100 per cent payment	4		
b 75 per cent payment	3		
c 50 per cent payment	2		
d 25 per cent payment	0		
4 Compensation			
a $3 million	3		
b $2 million	2		
c $1 million	1		
d no compensation	0		
5 Loan			
a $6 million loan	1		
b $4 million loan	3		
c $2 million loan	5		
d no loan	2		

TOTAL

Minimum needed for agreement is 12 points out of 20

Summary of points for the State Water Commission

Your name:

Did Irrigation Inc get the agreement it needed?
YES   NO

If 'yes', which parties agreed to the proposal?

State Water Commission   Didgeredoo Shire
State Environmental Council   WAC
Meatloaf Creek Irrigators

What was the agreement, which options were agreed to and how many points were generated?

Issue	Outcome	Points
1 Revegetation		
2 Water quality		
3 Payment		
4 Compensation		
5 Loan		

Total points

14 Stage 4: Action and Cooperation

All the players in the game

LEARNING MATERIALS

- Leadership approaches
- Leadership style questions
- Sharing leadership
- An open forum

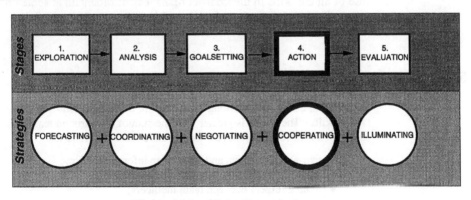

Figure 14.1 All the players in the game

AIMS AND OBJECTIVES

The exercises and self-tests illustrate Stage 4 of managing environmental change. They are intended to provide:

- practice in leadership skills needed to form a cooperative team;
- opportunities to identify and work with a wide range of different leadership styles;
- experience in negotiating on behalf of a team; and
- diagnosis of your own leadership style.

This session examines what it means to have an effective, cooperative team. Such teams require every member to share both the opportunities and the risks involved in the teamwork. Good teams can emerge in the most unlikely places and under the most adverse conditions, but there seem to be some general rules. A good team contains a range of styles and skills which have been worked into a cooperative system. These skills are made, and can be learnt. They are not skills that anyone is born with.

Accepting responsibility in a team is also accepting being a part of a power relationship. Leading from behind is one type of power – sharing responsibility equally between all members is yet another.

The need to manage environmental issues calls for teamwork in many work areas where teamwork may be quite unfamiliar. For example, environmental scientists, health professionals, town planners and dieticians may find themselves part of the same team when asked to deal with cities that are built on previously industrialised land. A fully-developed team would make use of all the skills in the Conflict Management Mountain at some time or other. The more skills the team has within its membership, the more stable it will be.

RISKS AND OPPORTUNITIES

Working in a team and working alone require different personal and interpersonal skills. 'The whole is greater than the sum of the parts' in teamwork, but that result is not easily achieved. Cooperative teamwork requires:

- two-way, top-down and bottom-up information flow;
- open communication between all members;
- mutual respect between team members;
- coordination across different problem-solving styles;
- commitment to team goals as well as personal goals;

- willingness to commit time and effort to support the team; and
- a capacity to negotiate constructively both within and on behalf of the team.

Two group management styles can look deceptively like cooperative team leadership at a distance, or for a very short time. One type of tightly knit group is the result of dictatorial management, that is, the team has an autocrat (leader), whose word is law. This type of team leadership may appear to work well to outsiders, but research shows that it is less productive and less creative than an cooperative team. At the other extreme is a group in which individual members never accept any form of leadership responsibility. The result is either anarchy with a dozen separate actors, or a vacuum in which a dictator inevitably emerges.

ACTIVITIES

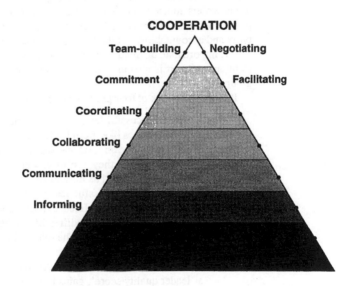

Figure 14.2 All the players in the game: activities

Theme: The Team Where Every Member is a Leader

The exercises in this session will allow each participant to:

- diagnose their own leadership style;
- test the style in practice;

283

- work with people with quite different styles; and
- evaluate different groupwork roles and styles.

MATERIALS

Leadership Approaches

Ian McAuley, Consultant, 24 Schlich Street, Yarralumla, Canberra, Australia

Main Leadership Theories

There is a confusing array of conflicting terminology relating to leadership. Many writers equate leadership with exercise of authority and with exercise of power. Baird, for example, talks about 'authority, power, delegation, responsibility and accountability' as the five basic aspects of leadership.

It is important to distinguish power from authority and distinguish leadership from authority. It is also unwise to assume that there is only one leader in a group.

It is useful to look at four prevailing approaches.

1 **Trait:** a successful leader has influence because of personal traits. We hear phrases like 'born leader', claims that so-and-so is a 'natural leader'. Some writings on leadership praise 'traits' such as expressive behaviour which they see as essentially 'feminine' or risk-taking which is stereotyped as masculine.

2 **Situational:** a successful leader has influence because he or she appears to match the fixed expectations of followers in a given situation. The leader modifies his or her behaviour to match the demands of the situation. The role of the 'leader' changes according to the demand of the group.

3 **Contingency:** a successful leader, based on something testable like a 'leader quality score', gains influence by picking the situation that corresponds favourably with his or her score. We are regarded as leadership only in cases where the situation matches our attributes. ('She's a great leader in troubled times, but not much in a stable organisation.')

4 **Transaction:** a successful leader earns influence by adjusting to the expectations of 'followers'. This is based on exchange of skills, resources or power.

The Limits of Conventional Wisdom

We can question some of the conventional wisdom more closely, particularly:

- the notion that power, authority and leadership necessarily reside in the one position; and
- the notion that there are 'leaders' and 'followers'.

Power, Authority and Leadership

First the distinction between power and authority. William Foote Whyte, a student of Homans, did a major study of what goes on in a restaurant – all the transactions, workflows, who is dependent on whom, etc. He found that power resided early in the production chain – with the waiters, although the authority lay with the manager. He made it clear that power and authority are separate entities, and there is no reason to expect they should always align.

Leadership and Authority

We are constantly tripping over writings and other materials which assume these are one and the same. The person with authority in a given setting may not be in the appropriate role to exercise leadership. Authority, because of restrictions associated with the position, and the expectations of the group, can actually constrain leadership from operating. Authority arises from a position, leadership does not.

It is therefore more useful to think about 'leadership' (an entity which is attached to a group, just as is cohesiveness), than 'leaders'. In this case the definition of 'leadership' is the capacity to mobilise groups to do work. Individuals can contribute to it, but ultimately it is a group resource.

Three dimensions, power, authority and leadership, give us eight binary combinations, with combinations as shown in Table 14.1

Table 14.1   The three dimensions of leadership

Power	Authority	Leadership	Example
Y	Y	Y	Traditional model of 'leader'
Y	Y	N	Air traffic controller, many technically competent managers
Y	N	Y	Boris Yeltsin, Nelson Mandela
Y	N	N	Lec Walenska, many union chiefs
N	Y	Y	Bill Hayden, good chairperson
N	Y	N	90-day wonder officer
N	N	Y	Mahatma Gandhi
N	N	N	Traditional model of 'follower'

We could imagine these as three overlapping Venn diagrams (Figure 14.3).

Figure 14.3   Power, authority and leadership

Lest we think such distinctions between authority and leadership are radical, it is worth going back to some very early notions – such as the Biblical story about Christ's temptation in the desert – when he was offered power and authority, which the devil knew would detract from his leadership. Similar stories exist in Buddhist scripture, and no doubt, in other traditional readings. We seem to have forgotten them in management circles.

Apart from clarifying the distinction between leadership and authority, we need to examine the pressures which tend to disempower people in authority positions. The expectation that the authority figure will do a group's work is destructive of group achievement – a form of 'work avoidance'. It leads to cycles of raised expectations and disillusionment.

Practical advice for those wishing to exercise leadership is:

- to help the group clarify problems, distinguish between immutable conditions and problems, and help it develop a practical vision;
- to listen;
- to withhold the temptation to solve the group's problems for them;
- to use allies, particularly if in an authority position;
- to beware of work avoidance – the techniques groups use to avoid working on hard issues.

As for the best leaders, the people do not notice their existence. The next best, the people honour and praise. The next the people fear, and the next the people hate. When the best leader's work is done, the people say 'we did it ourselves'   Lao-Tzu

Leaders and others

If leadership can be separated from authority, need we think of leaders and others? Or leaders and followers?

All the activities in this manual call on each individual's capacity to exercise leadership – in any group, and from any position. A useful contribution all can make is to identify and break patterns of 'work avoidance'. What are these pattens in some of the groups in which we operate? Who leads them? Who resolves them?

LEADERSHIP STYLE QUESTIONNAIRE

Directions: Reply to each item according to the way you would be likely act if you were a leader of a work group.

Items:		I would:		
1	Criticise poor work	Yes	Don't know	No
2	Most likely act as spokesperson of the group	Yes	Don't know	No
3	Encourage people to work overtime	Yes	Don't know	No
4	Do personal favours for group members	Yes	Don't know	No
5	Put most suggestions made by group members into operation	Yes	Don't know	No
6	Treat all group members as equal to self	Yes	Don't know	No
7	Work to a plan	Yes	Don't know	No
8	Make sure all group members are working as hard as they can	Yes	Don't know	No
9	Make sure all group members follow the rules	Yes	Don't know	No
10	Take time to listen to group members even if they want to talk about things other than work	Yes	Don't know	No
11	Explain all my actions	Yes	Don't know	No
12	Consult with all the group members before making a decision	Yes	Don't know	No
13	Decide what should be done and how it should be done	Yes	Don't know	No
14	Stress being better at the work than other groups	Yes	Don't know	No
15	Make sure everyone knows what is expected of them	Yes	Don't know	No
16	Be friendly and approachable	Yes	Don't know	No
17	Allow group members to do their work the way they think best	Yes	Don't know	No
18	Do everything to make group members feel at ease when talking with me	Yes	Don't know	No

---

Sorting

Work out your own scores on the three styles of leadership (Task, People and Laissez-Faire) by:

1 Placing a tick in the T-Style and P-Style columns below if you answered 'Yes' to the items shown. Then add the number of ticks to get a T-Style score and a P-Style score.
2 For an L-Style score, simply add all the 'Don't Know' answers.

T-Style	P-Style	L-Style
1_____	4_____	_____
2_____	5_____	_____
3_____	6_____	_____
7_____	10_____	_____
8_____	11_____	_____
9_____	12_____	_____
13_____	16_____	_____
14_____	17_____	_____
15_____	18_____	_____

TOTAL   **TOTAL**   **TOTAL**

Participants should not be asked to reveal their scores, unless they wish to discuss them.

This checklist of leadership style puts everyone, somewhat artificially into one of three styles:

1 **Task-centred:** Wishes to get the job done, and achieve the intended outcome; more interested in the product of the activity than in the process of achieving the product.
2 **People-centred:** As interested in the process by which the task is achieved as in the outcome itself; spends time and energy on the people involved in achieving the goal.
3 **Laissez-faire:** Allows the existing system to follow its course, without particular emphasis on either task or people.

Given that all these styles will be useful in some leadership conditions, the checklist is a way of identifying a preference or a prejudice towards a certain style.
Course members could discuss:

● the relationship between the three leadership styles and the needs of modern environmental management;
● the aspects of leadership reviewed in Ian McAuley's article and the leadership styles from the checklist; and
● ideal leadership characteristics in environmental management.

Sharing Leadership: Cooperating in a Team

Effective cooperative teams require that each member share the risks and the responsibilities. A good team contains a range of styles and skills which, working together, can contribute much more than the sum of the individuals' contributions.

Leadership Style (30 mins)

Participants respond to the set of trigger phrases in the leadership style inventory, and are allocated to one of the following leadership styles:

- task-centred
- people centred
- laissez-faire.

Each style is typical of certain roles and professions. Everyone uses all the styles at some time. Group members could discuss the appropriateness of each style in relation to trends in environmental management.

Leadership When It's Your Problem (TRIAD) (30 mins)
Leadership When It's Their Problem (TRIAD) (30 mins)

Sets of three people each with a different leadership style, meet to discuss what action they would take towards the proposed goal.

Within each sub-group, allocate labels of A, B, and C.

- A. has ten minutes to describe his/her action plan on a selected environmental issue.
- B. makes constructive and clarifying comments throughout, in order to assist A.
- C. observes the interchange, taking ten minutes to feed back (a) the style in which B is assisting A, and (b) the extent to which A is learning from B. C may find the attached observation sheet useful, and the triad may like to generate their own questions.

After 20 minutes, A moves to B, who moves to C, and C to A, and repeats the process.

After another 20 minutes, A takes C's task, B takes A's, and C takes B's.

Discussion (30 mins)

Triads (sets of three people) rejoin their group and discuss the value of different action strategies and different helping styles.

Group members should acknowledge:

- contributions of each member,
- insights into substantive issues,
- insights into conflict management, and
- insights into group functioning.

Clarifying Triad Observation Checklist

When you are observing the sender and receiver in your triad practising the skill of clarifying, use the following checklist. Apart from writing 'yes' or 'no' to the questions or recording the number of times an activity occurred, it is helpful for the later discussion that you will lead, to quickly note examples of behaviour displayed during the discussion between the sender and receiver.

- Is the receiver endeavouring to *gain an understanding* of the sender's ideas?

- Is the receiver *reflecting back* to the sender his understanding of what the sender means?

- Is the receiver *rewarding* the other's statements?

- Are the *general statements* being examined for specific meanings?

- Are *specific statements* being examined for general meanings?

An Open Forum: Listening to all the Players.

The Open Forum: Consulting with Stakeholders in Bangkok

Read the case study *Bangkok: Trouble with Traffic*, Chapter 8.

The increasing levels of public participation in Thai government and planning suggest one approach that may be effective in finding cooperative solutions to the city's traffic problems: the Open Forum.

In this exercise, the group will represent the people and groups of Bangkok: residents, government officials, traffic police, industrialists, small family businesses, poor people, slum dwellers, Buddhist monks, social reformers, environmental NGOs, and all of the other interest groups in the city. Note that the group needs to identify all of the social elements of Bangkok, not just those that have an obvious part to play in the management of the city's traffic. All groups are affected by Bangkok's traffic crisis, therefore all need to be involved in its resolution.

Having identified all of the sections of the community, the group should distribute roles to each participant. Some participants may have several roles. For instance, a radio announcer can also be a Buddhist monk. Presumably most people will be residents and transport users. Remember to include people who do not at first appear to be directly concerned with road traffic. For instance, people living along canals and the drivers and passengers of hired boats, are affected when people return to using water transport instead of roads.

The next stage is for every person to outline what they see as the problems with Bangkok's traffic, and the reasons for them. Issues may include the number of cars on the road, the efficiency of public transport, air pollution, the number of hours that people have to spend in traffic, the number of working hours lost, and the poor integration of freeways and other roads. Each person should pay particular attention to the issues that affect their character. At the same time, everybody should feel free to make comments about any aspect of traffic in Bangkok.

Some suggestions may appear contradictory – a driver may feel that the freeways are making traffic conditions worse; a planner may feel that the freeways are working well. Don't attempt to sort this out now. Stick to finding out what the issues are. Ensure that everybody has a chance to contribute, and that everyone feels that they have had a chance to contribute. No-one should ever contradict another person – what is visible as a problem to one person may be invisible to another. Now is not the time for sorting out whether a problem is or is not important. Individuals should be able to add to the suggestions of others, only so long as it does not impair the meaning of the original statement.

Do not rush this stage of identifying problems. Do not propose solutions or management options yet: these can wait until later. No solution can possibly be effective unless all of the problems, issues and interest groups have been identified.

Everybody should be quite clear on what other people mean when they say that something is a problem. This will involve a great deal of talk, either between individuals or within the group as a whole. If people do not understand why something is a problem, or even that something *is* a problem, they cannot possibly solve it. Good understanding requires good communication and a good deal of effort. Very often, two people who say they have the same problem may actually be looking at quite different problems, and without discussing what each means, they will never realise the difference resulting in misunderstandings. Take the time to talk.

Of great importance is the unequal power between the different interests in the issue. Special support will be needed for the people who do not often have a voice, such as women raising families in slums next to busy expressways; and people living along the waterways polluted by the traffic. Special controls, such as committee procedures or a very firm chairperson will be needed to ensure that those used to speaking and arguing in public do not dominate proceedings. The planner usually has a stronger voice that the car driver in such meetings. The police superintendent may consider that the Buddhist monk speaking for residents has no right to an opinion.

When all of the issues facing Bangkok have been identified, the group may want to organise issues together into several categories. Alternatively, the group may wish to leave them as separate, self-contained issues. Or the group may see all of the issues as interrelated and not easily categorisable or sub-divisible.

Having identified what all of the problems and issues are, it is important for the group to identify why all of the points raised *are* problems and issues. In particular, it is important to examine issues that appear to contradict one another (see Table 14.2 for example).

In Table 14.2, although on the surface the issues appear contradictory, both groups share the same concern about getting people into and out of the city quickly and easily. By attacking the *person*, the problem would never be solved. By examining the *issues,* and looking at them rather than the other person as the problem, both parties may find considerable common ground. Trust is an important part of this process. Some people will feel uncomfortable in this situation. The rest of the group should be willing to listen to a participant without bias. No-one should feel that they are being attacked for their views. Concentrate on the issues and problems; also, stick to facts, reasons and things that people are sure about – avoid supposition and guesswork. If anyone is unsure of anything at any point, they should feel free to ask for information or clarification. No-one should feel they have to second-guess what other people are thinking.

Table 14.2 Examination of contradictory issues

Driver	Planner
The issue	
There are too many expressways	There are not enough expressways
The reasons	
They encourage more cars onto the road, making the traffic worse	If there were more expressways, the roads would be able to cope with the increase in car numbers.
They are badly integrated with other roads	Expressways take the traffic load off other, less well-designed roads in the city.
Traffic only moves on them at 20 km/h – hardly 'express'	Traffic would move faster if more expressways were built
The expressways do not make it any easier for me to get into the city	Without more expressways, it will not be any easier to get into the city.

It is important to remember that the problem in front of the group is to solve Bangkok's traffic problems. The aim is not to 'beat' the other participants, but to beat the problem. In this sense, everyone is on the same side and everybody needs to help and support each other. Trust, openness, and a willingness to contribute and to listen are all essential to this process. Someone that is not willing to contribute cannot expect to be fully involved in the solution. Someone that is not prepared to listen will not understand the full nature of the problem and so will not be able to contribute fully to a successful solution.

With a full understanding of what the issues are and why they are problems, the group is now in a position to propose solutions to manage Bangkok's traffic. Creative answers are what are needed, and everyone should be encouraged to think freely and laterally. Explore areas of common interest and build on them.

At the initial stages, the more ideas, the better – no matter how crazy they seem at the time. Pruning can come later. People need to feel that they can contribute anything freely and without the thought that they may be humiliated, ignored or laughed at for their suggestions. As this phase progresses, people will start to feed off other people's ideas. This is an entirely natural process, and people should not feel too possessive about 'their' ideas. Begin to draw ideas together, consolidate them and flesh them out. This way, the volume of suggestions will decrease, but not the range of solutions. Rough ideas will begin to acquire the polish of a finished plan.

The process of developing ideas is evolutionary. Ideas are born, grow and may eventually be subsumed or replaced by other ideas. Let ideas and suggestions die if they no longer serve a useful purpose – but don't let ideas go if they are not being adequately replaced. Some ideas will not fit into any plan. An idea that does not fit neatly into a plan is not necessarily a bad idea. Proposers may want to withdraw ideas in the face of others' ideas. The group may decide by common consensus that an idea is no longer applicable – but no-one should feel that their ideas are being ignored, or worse, demolished.

As the plan takes on a firm form, establish objective criteria on how action will proceed, and how to determine how effective actions are. Establish criteria for success. Make firm objective statements on what is to happen. Statements like 'the Traffic Police will improve traffic flow' are no good. They do not say what is going to be done, and how to tell if it has actually been achieved. Much better to issue a statement along the lines of:

In the next twelve months, the Traffic Police will:

- recruit and train five hundred new members to direct traffic at busy intersections with the aim of improving driver behaviour;
- order and supervise the removal of stalls along Anuchat and Poungsomlee Roads;
- establish links with the Office of Town Planning to ensure that traffic around construction sites is minimised.

And so on. It is clear what the police will do, when they will do it, how they will do it, and what the final outcome will be. And there is a simple way to ensure that the programme has been successful.

It is important to state in the plan how long it is to run, when an assessment of the plan's performance is to be made, and who is going to do the assessing.

Having established a plan of action, it is now worth considering the future. Is everyone going to be involved forever more? Is a committee going to be established to watch over the plan? How many people are going to be involved? Do they want to be involved?

The exercise should finish with:

1   a clear idea of what the issues facing Bangkok are;
2   a clear understanding of why the issues are of concern to people;
3   a plan to manage Bangkok's traffic;
4   a clear understanding of who is going to do what and when it is going to happen.

15   Stage 5: Evaluation and Illumination

All the players know the game

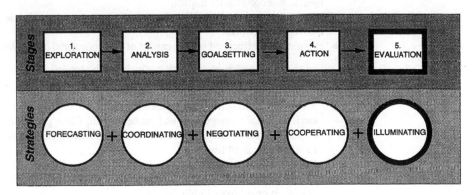

Figure 15.1 All the players know the game

AIMS AND OBJECTIVES

- To evaluate the different elements of an environmental issue of importance to community stakeholders, technical experts and professions.
- To develop skills in combining evidence from the qualitative and quantitative sources of value in practical decision-making.
- To accept uncertainty and a range of different options as the outcome of an evaluation.
- To consider evaluation as an illuminative rather than judgmental process.

In this session, participants work on the results of previous sessions, or on an environmental project with which they are familiar. They will be asked to identify the range of clients interested in the results, and the feedback mechanisms to be used during the evaluation.

Conflict management lies in achieving mutual understanding of, and assistance with, acceptance of the results of the evaluation.

Stage 5 of the environmental management process is the review process: illuminating both the intended and unintended outcomes of an environmental program or strategy. The traditional method of environmental evaluation – the Environmental Impact Assessment (EIA) – has been found wanting, on the grounds that it has:

- been a one-off review, undertaken either before or after the event, and not as an integral part of the process;
- examined the physical environment only, and not included the social agenda and community impact;
- provided an uncoordinated list of specialist reports, without discussing the implications; and
- examined the effect and not the causes of environmental deterioration.

These are encouraging trends in evaluation which can be looked for to supply the basis for future State of the Environment Monitoring.

1   There are approaches to evaluation which see it as illuminating environmental management and process itself. Clear principles are established for all players monitored, and the results communicated to all players at all times.

2   Environmental management is being perceived as managing social, as well as environmental, change. Models for studying social change can be co-opted from Public Health.

3 Many organisations, from the OECD to the UNCSD to national departments of environment, are developing principles for evaluation of environmental management.

RISKS AND OPPORTUNITIES

A risk for even the very best evaluation is that the findings will be ignored. Those who most need to hear the results are often those least willing to hear them, They may reply with 'We knew that already – there were no surprises', or they *may not listen* at all. Or those being evaluated may hear only one predetermined answer, reflecting a hidden agenda which has nothing to do with the environmental issue at stake. A means of avoiding either fate is to integrate the environmental impact evaluation with the social currents of the time by including questions on the full social context in which the impact occurs.

There is always a need to identify unintended as well as intended outcomes: future changes as well as the end result. Public health practitioners use a framework capable of evaluating social change. This framework illuminates the policy, community and service provision aspects of environmental impacts. The five dimensions to be examined here are as follows:

1 Integration of environmental policies from all sectors, as recommended in integrated catchment management, whole farm plans or primary environmental care.
2 Quality of natural and social environments, based on a comprehensive set of environmental indicators, such as the Valdez Principles, described later in this chapter, or the OECD Impact–Condition–Response model.
3 Strength of community action, monitoring community contributions to the preparation, implementation and evaluation of environmental policies.
4 Levels of individual skills in environmental decision-making and conflict management.
5 Reorientation of agencies and services toward preventing damage, rather than merely providing remedial services after the damage is done.

ACTIVITIES

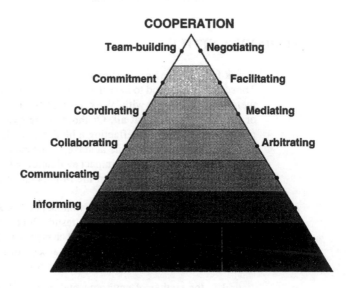

Figure 15.2   All the players know the game: activities

Theme: Sharing the Light

Brainstorming between participants. Develop a group list.

- Identifying the uses of evaluation.
- Selecting evaluation criteria – Nominal Group Process.
- Process evaluation – Action Research.
- Criteria for environmental evaluation.

The skills learnt from these exercises will include:

- methods for illuminative evaluation;
- the multiple purposes of evaluation;
- process, summative and reflective evaluation; and
- principles of environmental evaluation.

MATERIALS

Evaluation as Illumination*

Definition of the Problems

In an illuminative evaluation, the general areas to be explored begin to be defined at the first stage – the negotiation. At this point the evaluator has extended contact with those commissioning the study (along with others who are stakeholders in the enterprise) and tries to establish what questions and problems are of most concern, and how the study can address them to best effect. Since many a good intention is jeopardised by trivial misunderstanding, numerous (well over 20) questions of a practical nature are also dealt with at this stage (eg What kind of report? Who will see the draft? What in-house facilities are available? How is the study to be introduced to participants?)

Illuminative evaluations are not designed in advance with a set of fixed questions or tests to be made. They are far more exploratory in nature, with the problems-to-be-studied being identified through intensive familiarisation with the issues and character of the programme. Only by appraising the full organisational, human, and political complexity of an environmental management issue can the evaluator fully comprehend the nature of questions on people's minds and what he or she can accomplish by various investigative probes.

The approach has as much in common with consulting as it has with research. Yet, unlike consulting, the outsider is not aiming to proffer prescriptions, recommendations or judgements as such. Rather, by appropriate study and reporting, he or she is trying to increase communal awareness and bring local as well as wider-scale policy questions into sharper focus. The evaluation provides, within a single analysis, information and comment (including many different persons 'evaluations') that can serve to promote discussions among those concerned with decisions concerning the system studied.

A study will often bring issues that have hitherto been only part of the 'background' of a setting, known at some level but effectively disregarded (eg. 'low staff morale') to the foreground of attention.

By drawing attention to staff morale, pulling together a variety of individual interpretations about why it is low, documenting how widespread it is, and showing how it has wider consequences, the evaluation study can have a profound effect on discussions of policy and practice. Our conviction is that the

---

* For further reading see Parlett, M (1978) 'Illuminative evaluation' in Reason, P and Rohan, J (eds), *Human Inquiry* Wiley, New York

conventional definitions of the evaluator's role are self-defeating and that over-concentration on 'impacts' and 'testing outcomes against intentions' can lead to the evaluator having his or her report promptly shelved.

To sum up, the subjects addressed in each study relate closely to concerns expressed: there are no fixed evaluation procedures and the exact purposes of each study are unique to that setting and to the particular policy discussions into which the report will be fed.

The Method

Evaluation is thus conceived in broad terms – it is the study of an organisation or decision-making process undertaken in such a way that it contributes to decision-making and review of policy. Such studies often include review of the purported merits, problems, advantages, and negative side effects of the program's present policies but only as part or a more extended organisational analysis.

From the foregoing it is obvious that studies have to be custom-built – the initial strategies being chosen in accord with the complicated contract or agreement that emerges from negotiation. Moreover, each study has to be open-ended enough for the evaluator to home in on critical emerging issues.

Problems identified and discussed at the negotiation stage may correspond to those finally addressed, but not invariably so. The exact areas discussed at the reporting stage are those judged by the evaluator to be significant issues – in the light of what has been found out: significant to the participants themselves (eg a common difficulty but not one publicly admitted); or significant for explaining certain phenomena (eg seemingly trivial procedures that alienate substantial numbers of people); or significant by virtue of being central to the concerns of one or more critical audiences for the study (eg the cost implications of proposed changes in the programme).

The flexibility called for in a custom-built design requires an extended range and choice of techniques to be used. They are chosen to fit the questions, opportunities and restrictions that a particular investigation poses: problems dictate methods rather than methods dictating problems. There has to be an assessment in each study of what methods will best serve investigative needs, with due regard to the time and other resources they consume; and also what will be responded to enthusiastically by those contributing information. Illuminative evaluations rely extensively on interviews and observing in the field, along with analyses of documents collected and short questionnaires, often open-ended in structure. In addition, the study of stored records (eg admissions data, test scores, costs, numbers of students pursuing different options) often forms an integral part of an in-depth investigation.

Using different techniques in parallel also provides for internal checks. Each method has limitations and there is often an advantage in combining techniques and triangulating on issues from different directions.

Evaluation as Business Ethics: The Valdez Principles

These principles have been developed by a wide consortium of industries in the United States. The consortium developed the principles by negotiation between themselves, as a guarantee that their businesses would not lead to another 'Valdez' incident, such as the vast oil spill in Alaska.

1 **Protection of the biosphere**: We will minimise and strive to eliminate the release of any pollutant that may cause environmental damage to the air, water, or earth or its inhabitants. We will safeguard habitats in rivers, lakes, wetlands, coastal zones and oceans and will minimise contributing to the greenhouse effect, depletion of the ozone layer, acid rain or smog.

2 **Sustainable use of natural resources**: We will make sustainable use of renewable natural resources, such as water, soils and forests. We will conserve nonrenewable natural resources through efficient use and careful planning. We will protect wildlife habitat, open spaces and 'wilderness', while preserving biodiversity.

3 **Reduction and disposal of waste**: We will minimise the creation of waste, especially hazardous waste, and wherever possible recycle materials. We will dispose of all wastes through safe and responsible methods.

4 **Wise use of energy**: We will make every effort to use environmentally safe and sustainable energy sources to meet our needs. We will invest in improved energy efficiency and conservation in our operations. We will maximise the energy efficiency of products we produce and sell.

5 **Risk reduction**: We will minimise the environmental, health and safety risks to our employees and the communities in which we operate by employing safe technologies and operating procedures and by being constantly prepared for emergencies.

6 **Marketing of safe products and services**: We will sell products or services that minimise adverse environmental impacts and that are safe as consumers commonly use them. We will inform consumers of the environmental impacts of our products or services.

7 **Damage compensation**: We will take responsibility for any harm we cause to the environment by making every effort to fully restore the

301

environment and to compensate those persons who are adversely affected.

8   **Disclosure**: We will disclose to our employees and to the public, incidents relating to our operations that cause environmental harm or pose health or safety hazards. We will disclose potential environmental, health or safety hazards posed by our operations, and we will not take any action against employees who report any condition that creates a danger to the environment or poses health and safety hazards.

9   **Environmental directors and managers**: We will commit management resources to implement the Valdez Principles, to monitor and report upon our implementation efforts, and to sustain a process to ensure that the Board of Directors and Chief Executive Officers are kept informed of and are fully responsible for all environmental matters. We will establish a committee of the Board of Directors with responsibility for environmental affairs. At least one member of the Board of Directors will be a person qualified to represent environmental interests to come before the company.

10 **Assessment and annual audit**: We will conduct and make public an annual self-evaluation of our progress in implementing these Principles and in complying with applicable laws and regulations throughout our worldwide operations. We will work toward the timely creation of independent environmental audit procedures which we will complete annually and make available to the public.

Selecting Evaluation Criteria: The Nominal Group Process

The nominal group process is a method for assessing community goals for perceptions of an activity in a way that overcomes many of the traditional problems of unequal representation of opinions and permits a common perception to develop. 'Nominal' refers to a set of items listed in order of preference, and 'group' to the use of group members to agree on the order. The method consists of a series of small-group procedures designed to compensate for the usual inequalities of social power that emerge in most planning meetings. Those who use the method should keep in mind that its purpose is to identify and rank problems, not to solve them.

The method is effective for generating ideas and getting equal participation from group members. It is not a means of clarifying values, nor is it a decision-making strategy. It is a method for arriving at informed agreement as to priorities. The method works as follows:

Arrange the Participants into Groups of Six to Seven Members

It is important that the size does not exceed seven in order to allow for appropriate interaction. Those selected as participants should be representative of, and knowledgeable about, the community or task in question.

Pose a Single Question to the Group, Summarising the Issue

It is best if the question can be in writing on a blackboard, flip-chart, or hand-out sheets. The question should be generated following consideration of

1   the objective of the meeting;
2   examples of the type of items sought;
3   the development of alternative questions; and
4   the pilot-testing of alternative questions with a sample group.

Examples of the type of question are: 'What do you consider to be the major purpose of this evaluation?', 'Of all the possible avenues for action on the environment, which do you consider were the most valuable?', 'Of all the topics we could include in this programme, which do you consider the most central?'.

Have the Participants of each Small Group Write Down their Responses

Sheets of paper with the question listed at the top can be given out; this provides an easy reference point for the group members. However, if this is not possible, writing the question down on a blackboard, flip-chart or overhead projector will suffice. Although the actual amount of time necessary to complete this assignment will vary depending upon the particular question which is posed, an approximate amount of time would be 15 minutes. It is important that the group proceed in absolute silence (this is the responsibility of the course coordinator). Such an approach enables the group to reflect carefully upon their ideas, and to be involved in a competition-free atmosphere where premature decisions do not have to be made. Have the participants of each small group write down their responses.

The Co-ordinator Elicits Individual Responses

First, one participant is asked to give their most important (to them) single response, the next gives their single response, and this continues until each participant has contributed a single response. As the responses are stated, they are written by the group leader on a blackboard or flip-chart, each item being given a number (1, 2, 3, etc). The same process is repeated for a second, then a third time and so on, until all contributions have been recorded. This procedure enables each group member to play a truly participating role. During this time, discussion is only permitted for points of clarification, not on the form, format, or meaning or value of a participant's response.

Clarify the Meaning of the Responses

Take time to inquire as to whether or not each response is clearly understood. Allow participants time to discuss what they meant by a particular response, the logic behind it, and even its relative importance. However, this is not the time for argumentation and lobbying. The group leader must direct the proceedings so that only clarification takes place.

Conduct a Preliminary Vote

From the original listing of responses on the blackboard or flip-chart, participants are directed to select a stated number of the items they consider the most important (eg out of the summary of 20 individual responses, each participant is to select and rank seven of them). This is accomplished by asking each participant to write each one of the statements selected on a separate card first, and then rank-ordering them. The topics can be pulled together either by group agreement on priorities, or numerically (seven points would be assigned to the least important). As a rule of thumb, group members can prioritise only five to nine items with some degree of reliability. The item with the largest numerical total represents the top priority issue.

Confirm the Vote

It is important to discuss the various explanations related to choosing. Discussion regarding the high vote-getters and low vote-getters may be of value. It may also be useful to redefine the meaning of selected items, to be certain that all participants are clear on their meaning. Identify and value the various perspectives and discuss how they can be illuminated in a review of the situation.

EVALUATION SURVEY FORM

The following is a guide to evaluation of a learning programme such as this one, or an environmental management strategy.

WORKSHOP EVALUATION

1 Group learning – the overall workshop/strategy

Did it meet your previous objectives? (copy attached)
Comments/Suggestions:

Did it provide unexpected outcomes?
Comments/Suggestions:

Did you find aspects disappointing?
Comments/Suggestions:

Will you find it valuable personally?
Comments/Suggestions:

Will it be valuable professionally?
Comments/Suggestions:

2 Workshop/strategy organisation

a) Administration

b) Food/resources

c) Venues

3 Course/strategy coordination
Communication between participants

a) Clarity of purpose
Comments/Suggestions:

b) Language
Comments/Suggestions:

c) Continuity of coordination
Comments/Suggestions:

d) Informal interaction (time and placing)
Comments/Suggestions:

e) Plenaries/group meetings (time and placing)
Comments/Suggestions:

4 Workshop content

a) Problem-solving framework
Comments/Suggestions:

b) Order of the activities
Comments/Suggestions:

c) Resources for the activities
Comments/Suggestions:

d) Case study information
Comments/Suggestions:

e) Case studies as a basis for management process
Comments/Suggestions:

f) Content of management materials
   – Introduction.
   – Exercise outlines

      – Basic resource materials
Comments/Suggestions:

g)   Organisation of this manual
      – Themes
      – Briefing materials
      – Exercise guidelines
      – Reference materials
   Comments/Suggestions:

h)   Relationship between content and process.
   Comments/Suggestions:

i)   What, for you, was the most valuable aspect of this course/strategy?
   Comments/Suggestions:

5 What three outcomes would you like to see from this course/strategy?

1

2

3

6 In your area, how would you see this used as an in-service curriculum?

INDIVIDUAL LEARNING

Session 1 Exploring
Valuable _____ Of no use
Comments/Suggestions:

Session 2 Analysing
Valuable _____ Of no use
Comments/Suggestions:

Session 3 Goal setting
Valuable _____ Of no use
Comments/Suggestions:

Session 4 Taking action
Valuable _____ Of no use
Comments/Suggestions:

Session 5 Evaluating
Valuable _____ Of no use
Comments/Suggestions:

Session 6 Individual learning
Valuable _____ Of no use
Comments/Suggestions:

Session 7 Organisational learning
Valuable _____ Of no use
Comments/Suggestions:

16 Individual Learning and Personal Change

What part do I play in changing the game?

LEARNING MATERIALS:

- Conflict resolution style revisited
- Leadership style questionnaire
- Individual plan for managing change
- Individual learning contract

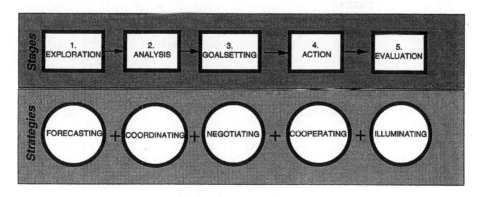

Figure 16.1   What part do I play in changing the game?

AIMS AND OBJECTIVES

The aim of this session is to pull together all the individual learning for each participant, so that they can evaluate the programme:

- for each participant's own learning; and
- for the understanding and application of conflict management.

Sessions 1–5 contain self assessment exercises, during which participants are able to determine their own style of conflict management, group leadership style, negotiation methods and problem-solving style. It is also hoped that participants will offer to share any skill that they have brought with them or insights that they have gained.

RISKS AND OPPORTUNITIES

The Adult Learning Cycle

The adult learning cycle assumes that, as adults, the workshop participants already have considerable expertise in environmental conflict management. New ideas and skills must be compatible with existing knowledge, or they will be, for all practical purposes, useless to the learner. The principles of adult learning as shown in Figure 16.2 have been applied in the organisation of the case study materials and in designing the practical exercises of this volume.

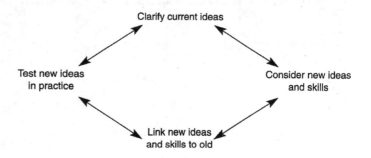

Figure 16.2   The principles of adult learning

ACTIVITIES

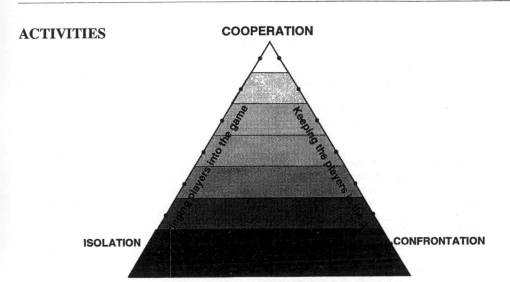

Figure 16.3   Bringing and keeping the players in the game

Theme: For Personal Change, Personal Style

- Review of personal style of conflict resolution
- Checklist of leadership style
- Individual management protocol and learning contract

MATERIALS

Conflict Resolution Style Revisited

Conflict Strategy

> The proverbs listed below can be thought of as some of the different possibilities for resolving conflicts. These proverbs reflect traditional wisdom for resolving conflicts. Read each of the proverbs carefully. Using the scale given below, indicate how typical each proverb is of your actions in a conflict.

CONFLICT RESOLUTION

5 – very typical of the way I act in a conflict
4 – frequently typical of the way I act in a conflict
3 – sometimes typical of the way I act in a conflict
2 – seldom typical of the way I act in a conflict
1 – never typical of the way I act in a conflict

		Score
1	Use soft words to win hard hearts	[ ]
2	Come now and let us reason together	[ ]
3	For your arguments to have weight, argue loudly and forcefully	[ ]
4	You scratch my back, I'll scratch yours	[ ]
5	The best way to handle conflicts is to avoid them	[ ]
6	When one hits you with a stone, hit them with a piece of cotton	[ ]
7	A question must be decided by knowledge and not numbers if it is to be a right decision	[ ]
8	If you cannot make a person think as you do make them do as you think	[ ]
9	Better half a loaf of bread than no bread at all	[ ]
10	If someone is ready to quarrel, they aren't worth knowing	[ ]
11	Smooth words make smooth ways	[ ]
12	By digging and digging, the truth is discovered	[ ]
13	He who fights and runs away lives to fight another day (once you strike, withdraw so that you can strike again)	[ ]
14	A fair exchange brings no quarrel	[ ]
15	There is nothing so important that you have to fight for it	[ ]
16	Kill your enemies with kindness	[ ]
17	Seek until you find, and you'll never lose your labour	[ ]
18	Might overcomes right	[ ]
19	Tit for tat is fair play	[ ]
20	Avoid quarrelsome people – they only make life miserable	[ ]

Check your results against your record sheet from the introductory session. Discuss whether any of the team have a different understanding of conflict and its management from the start of the course.

CONFLICT RESOLUTION SCORE SHEET

Conflict strategy

	Proverb no	Score
Avoiding	5	_____
	10	_____
	15	_____
	20	_____
		Total _____
Forcing	3	_____
	8	_____
	13	_____
	18	_____
		Total _____
Smoothing	1	_____
	6	_____
	11	_____
	16	_____
		Total _____
Compromising	4	_____
	9	_____
	14	_____
	19	_____
		Total _____
Problem solving	2	_____
	7	_____
	12	_____
	17	_____
		Total _____

Note: The higher the total score for each conflict strategy the more frequently you tend to use this strategy. The lower the total score for each conflict strategy, the less frequently you tend to use this strategy.
For interpretation of results, use the same diagram as in Chapter 10.

Discussion

Note: participants should not be asked to reveal their scores, although they may wish to do so themselves.

Calculating each person's score will take about 20 minutes. There should be at least 20–30 minutes of discussion about the implications of the personal scores. The following points should be made:

- all types of conflict management are appropriate at some point;
- everyone uses all five responses at some time;
- the ideal is to choose the appropriate response rather than fall into a familiar pattern.

Participants may like to discuss their own experiences with each conflict resolution style and types of environmental conflict for which each of the five types is most suitable.

LEADERSHIP STYLE QUESTIONNAIRE

Directions: Reply to each item according to the way you would probably act if you were a leader of a work group.

Items: **I would:**

1	Criticise poor work	Yes	Don't know	No
2	Most likely act as spokesperson of the group	Yes	Don't know	No
3	Encourage people to work overtime	Yes	Don't know	No
4	Do personal favours for group members	Yes	Don't know	No
5	Put most suggestions made by group members into operation	Yes	Don't know	No
6	Treat all group members as equal to oneself	Yes	Don't know	No
7	Work to a plan	Yes	Don't know	No
8	Make sure all group members are working as hard as they can	Yes	Don't know	No
9	Make sure all group members follow the rules	Yes	Don't know	No
10	Take time to listen to group members even if they want to talk about things other than work	Yes	Don't know	No
11	Explain all my actions	Yes	Don't know	No
12	Consult with all the group members before making a decision	Yes	Don't know	No
13	Decide what should be done and how it should be done	Yes	Don't know	No
14	Stress being better at the work than other groups	Yes	Don't know	No

---

15	Make sure everyone knows what is expected of them	Yes	Don't know	No
16	Be friendly and approachable	Yes	Don't know	No
17	Allow group members to do their work the way they think best	Yes	Don't know	No
18	Do everything to make group members feel at ease when talking with you	Yes	Don't know	No

Sorting

Work out your own scores on the three styles of leadership (Task, People and Laissez-Faire) by:

1   Placing a tick in the T-Style and P-Style columns below if you answered 'Yes' to the items shown. Then add the number of ticks to get a T-Style score and a P-Style score.

2   For an L-Style score, simply add all the 'Don't Know' answers

T-Style	P-Style	L-Style
1_____	4_____	_____
2_____	5_____	_____
3_____	6_____	_____
7_____	10_____	_____
8_____	11_____	_____
9_____	12_____	_____
13_____	16_____	_____
14_____	17_____	_____
15_____	18_____	_____
TOTAL	**TOTAL**	**TOTAL**

Individual Plan For Managing Change

1 **Identify a problem** or issue that is a high priority for you in the immediate future (and/or present):

2 **The change you require is:**

3 **The pattern of change is:**
   - Exploring
   - Analysing
   - Goal-setting
   - Acting
   - Evaluating

   Players

   - political/policy makers
   - technical/scientific (experts)
   - community group representatives
   - individual stakeholders
   - agencies, services

4 **Management of the change will be through:**
   - Conflict management style strategy
   - Leadership strategy
   - Communication
   - Teamwork

5 **Risks:**

6 **Opportunities:**

Individual Learning Contract

Name:

The task you set yourself:

Ways of completing the task:

Resources available:

Timetable:

Evaluation criteria: how you will know you have done it:

17 Organisational Learning and Social Change

Changing the game

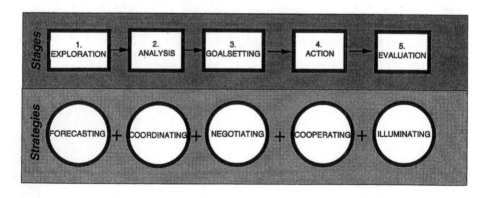

Figure 17.1 Changing the game

AIM AND OBJECTIVES

The aim of this final session is to identify the support and barriers which participants will meet when they return to their organisations or home-bases, and wish to implement any techniques that they have learnt from the workshop.

This session will:

- assist participants to apply their learning to their own environment;
- provide participants with a context in which to apply their learning in their usual roles; and
- help participants to set realistic future goals.

The session will also evaluate the workshop as a whole.

The Adult Learning Cycle in the previous session shows individual learning building on knowledge and experience. The new learning will only be incorporated in a person's repertoire if it:

- is compatible with previous experience (this may mean that previous experience has to be reviewed); and
- has been put into effective practice.

Once someone has the new knowledge and skills, they will want to put them into effect in their usual work or home environment. At this point, a whole new set of potential conflicts arises.

- the person has changed but the organisation has not;
- organisations are built upon stability not change, and the forces take a lot to build and even more to change; and
- organisational learning is quite different from individual learning. It depends upon
  - a pull for change from above
  - a push for change from below
  - a change in the critical mass of behaviour.

RISKS AND OPPORTUNITIES

Workshop members will need to identify the factors which could either enhance or inhibit the incorporation of their learning in the programme into their own organisations.

When participants return to their workplaces, they may find that introducing new methods of managing conflict is like hitting a brick wall. On the other hand, they may find that they feel lost and are unable to find the old place they left or a new niche in the organisation.

Each participant should choose a project of conflict management which they want to test when they return to their work.

Discussion

Group members might discuss their proposed conflict management project and their preferred role as:

- a revolutionary
- a catalyst
- an innovator
- a critic
- an observer
- a protector of existing systems
- a bulwark against change.

ACTIVITIES

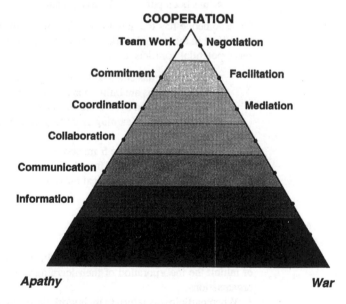

Figure 17.2 Learning the steps to cooperation

Theme for Organisational Change: 'Driving and Restraining Forces'

Participants will be able to share insights and techniques for which they have found effective practical applications in their own fields.

MATERIALS

Structural Change: A Paradigm Shift for Environmental Management

Peter H Cock

The focus of environmental management must be to shift from the need to manage the natural environment, towards increasing our capacity to manage the human environment, That is move from the victims to their causes. As practised through use of this manual, this involves a mix of levels of decision-making, mechanisms of management as well as personal attributes.

What is being asked of effective environmental conflict managers involves developing a generalist wisdom that works with paradox and dialectic as much as with specialists' skills. The environmental science becomes subsidiary to the art of management. The task of science is to provide data and the art is to work creatively with the dialectic between the mix of factors and issues; to discover and apply outcomes that are a synergy between the factors and interests involved. This involves taking a stance that is humble with nature and assertively radical with ourselves.

Policy and management now has to encompass multi-objectives, multi-strategies and tactics in order to work effectively with the complexities of the ecosystem and its multi-species populations. Clearly one needs to engage in strategic thinking and planning, to gather data, clarify preferred options, lobby support, organise resources. We need to address the structural issues that shape us and be capable of not making a decision. Until we know what we don't know, we need to ask how we can reduce the planet's dependence on us for its survival.

Our accepted and treasured social structures and processes have now to be reexamined in terms of their environmental outcomes. This includes reviewing the environmental performance of democratic processes. For example, unlike the command economy, democratic processes that reach throughout the society are a vital means for the mobilisation and recognition of diversity. A variety of inputs into the decision-making process enriches its complexity and the legitimacy of decisions.

A risk is that when there are too few, representing too narrow a range of interests, the public/environmental interests can be sacrificed in order to come up with a compromise between the participating interests. How to ensure as wide a range of interests are involved as possible, but in a way that maximises the probabilities that wise decisions are made? The more people/interests are involved, the more difficult it is to manage outcomes. Are the outcomes more likely to be benign for people and for environment?

A risk is that we hand over the environment to the professionals, a process that has failed in the welfare area. This approach means that one is inevitably locked into crisis management that grows because the structural causes are never addressed. It is in the interests of the profession for prevention to be a backwater to investment in symptom suppression and the treatment of the growing stream of victims. At its peak a monopoly is created where only experts can act, only paid workers can do the work – a government/union partnership to keep out concerned citizens and more importantly public/environmental interests.

As this conflict management handbook illustrates, there are social structures and processes for bringing together interests, disciplines, professions and institutions to enrich the decision-making process, to develop ways to get to 'yes', to generate a synergy that transcends previous conflicts, one-dimensionality and to sustain the capacity to say 'no'. These are all vital pathways to enhance the contribution human decision-making and management needs to play to achieve the reestablishment of sustainable ecological processes. They are vital to facing the diversity within our species and working to mobilising it for our common good. Our self-management from a species viewpoint is vital but needs to be seen in tandem with reducing the need to manage the environment.

In evaluating our performance and the tools we have used to manage environmental conflicts we need to be mindful that we cannot do it all. Increasing our skills as outlined and practised through this manual is a vital tool to working with the environmental complexity we face. Effective decision-making and management is dependent on an array of factors beyond individual, group, interest, organisation, and strategy, although each is necessary. No matter how capable we are, we need to evaluate conflict management tools in terms of their capacity to minimise the need to manage conflict – prevention has to be part of the treatment. This can only be achieved by making decisions that get us off the decision-making and management escalator.

This escalator is driven by:

- increasing social complexity;
- growing globalisation;
- the increasing material clutter of superficial material choices;
- increasing technological dependency and complexity;
- the decisions and management needed to compensate for the destruction of the capacity of Gaia to manage.

One major risk is that we are heading towards the idea of the super manager, the environmental manager equipped with open systems thinking, multi-skilled, using interdisciplinary team action. No matter how well resourced one's organised interest, and the skills of one's advocacy, there are still unpredictable outcomes. You can not do it all yourself, in your team or even with the aid of other interests and organisations.

Learning to live comfortably with our limited power is one of the keys to being effective as a participant in environmental decision-making and management. To manage the human environment effectively we will need to reduce Gaia's dependence on us, otherwise we are in danger of being overwhelmed by the demands for decisions and management. Both require plenty of time and a limited number of demanded decisions. Resolutions that have elasticity, so that if a synthesis is not achieved or a wrong decision is made then there isn't a significant determinant to Gaia or to human interests.

The development of conflict resolution and management processes require a supportive cultural and social structural context to be effective. For these structural conditions to be met requires reducing time scarcity and the number and significance of decisions. This can be achieved by backing away from the eco and cultural abyss in two basic ways:

1   through developing in partnership with the planet and its galaxy;
2   through cultural and social structural transformation that provides a community base to social life, thereby reducing the burden of state and individual decision-making.

Our task is to develop an approach to decision-making and management that is geared to minimise the need to intervene. For example, the larger the size and ecological complexity of parks, the more ecologically self-managing they can be. Being able to minimise our need to manage is dependent on the redevelopment of a human culture that radically needs less from its environment.

Force-field Analysis

Enhancing and Inhibiting Factors for Change (180 mins)

Participants consider the five stages of environmental management in relation to their own organisation. In small groups of seven to ten, the group is asked to build a collective model of organisational change. On a whiteboard or large sheet of paper, everyone brainstorms all of the enhancing factors (above the line) and all of the inhibiting factors they can think of (below the line) – see Figure 17.3. Each group member then identifies the factors which apply to their particular organisation and prepares an organisational profile. Members form subgroups of, say, three to assist each to identify the changes needed to support the enhancing factors, and reduce the inhibiting factors. The subgroups could agree to work together (as students or colleagues) as they bring about change in their respective organisations: changes difficult and unsettling; dealing with conflict produces stress for the manager and external support is very important to maintain perspective and staying power.

Forces inhibiting change

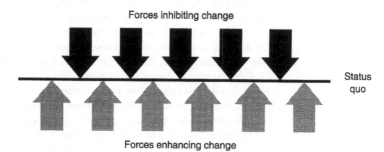

Status quo

Forces enhancing change

Figure 17.3   Factors for change

Triad Support Groups (120 mins)

The group reconvenes, in order to review the ideas and skills covered in the learning programme, to identify where these would be useful in returning to their own organisation, and to identify which skills they most need to develop further and how they plan to do this.

Discuss in threes on opportunities to apply the learning from the forcefield analysis to the projects chosen by each individual. The trio should use the feedback style described for Chapter 14 to help each other clarify the potential, and the action plan for each project. All the tools covered in the learning programme should be considered.

Case Studies of Organisational Change

In the Tiger's Mouth: Putting Course Principles into Practice

In July 1991, a four day workshop on Conflict Management in Environmental Decision-Making was held at the Centre for Resource and Environmental Studies (CRES) at the Australian National University, in Canberra. Participants developed ten projects to test the workshop themes and training exercises by applying them to real-world issues. Putting theory into practice has been described as 'putting one's head into the tiger's mouth'. The following environmental management projects have put the strategies from the learning modules the into practice.

Table 17.1 Tools for conflict management

Problem-solving stage	Skills	Exercises
1 Forecasting	Vision	Search conferences using a strategic framework Vision workshops Futures Wheel
2 Coordination	Group management	Stakeholder analysis Teambuilding Community consultation Consensus conferences Committee procedures
3 Negotiation	Reaching agreement	Nominal group processes Harvard Negotiation Process Community needs analysis
4 Cooperation	Teamwork	Partnership, effective leadership Development teamwork Issues management Management styles
5 Illumination	Evaluating	Linking factual, social, ethical and personal aspects of the evidence Choosing performance indicators
6 Personal learning	Learning diagnosis	Communication style Learning style Skill sharing
7 Organisational learning	Managing change	Evaluating personal and organisational change Force-field analysis Shared group learning

The objectives of the projects were to:

- develop a range of case studies of different conflict management applications;
- test the participants' ideas and personal skills in real situations;
- achieve mutual advantage solutions to environmental issues; and
- develop materials for a conflict management manual.

Questions which were asked during the projects were:

- the applicability of the manual materials;
- situations that go beyond the manual materials;
- identification and management of conflict triggers;
- differences between public and private sectors;
- differences in conflict management expectations across different professional fields.

The workshop staff at CRES provided support for the workshop participants to develop their projects. This support was in the form of:

- a short workshop for project team leaders (in September 1991);
- informal support for each participant on request;
- reading and resource materials relevant to each topic;
- constructive critical discussion of each issue and project;
- teleconferences at the request of any participant;
- a formal letter of support for participants' projects addressed to their organisation;
- a review of each participant's project on request;
- a commitment to write an outline of each project, to be edited and used with the permission of the participants, in this Conflict Management Manual.

Projects

Integrated catchment plan

Viv Sinnamon

The aim of this project was to develop an Integrated Catchment Management Plan for the Mitchell River in Far North Queensland. The Mitchell is Australia's second largest river and drains most of Cape York. In the catchment are properties grazing cattle, mining developments, Aboriginal communities and several European settlements. The Mitchell flows into the Gulf of Carpentaria, and its estuary is an integral part of the fish and prawn breeding grounds. Fishing and trawling are growing industries in the Gulf. Issues to be addressed in the development of the Catchment Plan include water supply, water quality, preservation of the catchment area, and development within the watershed boundaries.

Specific aims of the project are:

- to apply the conflict management skills from the programme in water resource management; and
- to apply conflict management skills in developing cooperative team work between Aboriginal and European communities, and State services.

Management of Kangaroo Island
Bruce McKenzie
Kangaroo Island is a mostly farming community, with large areas of untouched countryside. Tourism is a growing industry, but is placing strain on the fragile native environment of the island. The specific aim of the project is to explore and manage the issues for local government in dealing with the conflict between conservation and tourism development.

Environmental policy development
Richard Kenchington, Peter McVay
As Australia develops its plans for Ecologically Sustainable Development, attention is being drawn to the potential for conflict between different government agencies, and legislation aimed at managing and planning for environmental change. At the level of Federal Government alone there are dozens of interests: the Department of Arts, Sport, the Environment, Tourism and Territories, the Department of Finance, the Office of Local Government, the Department of Primary Industries and Energy, and the Resource Assessment Commission are only a few of the key players involved in Commonwealth planning. State and Local Governments have their role to play too, and have varying resources and opportunities to contribute. Over three hundred Acts of Federal and State Parliaments dealing with different aspects of the environment exist. These issues, coupled with undefined political positions, are a clear opportunity for conflict.

Two projects were begun in this area of environmental policy. The first was aimed at applying Conflict Management techniques in the development of environmental policy. The aim of the second project was to develop decision-making models for resource development and environmental management, bearing all of the above issues in mind.

Transport
Kerry Willis
As the size and nature of Australian urban centres change, so to does the pattern of transport use and access. An issue of direct importance to residents in many

327

Australian towns and cities is the heavy transport use of residential areas. Various aspects of the issue include the siting or resiting of major transport routes, acceptable traffic levels, traffic flow, air and noise pollution, community input into planning, and options for redressing current difficulties. Specific aims of this project were:

- to prepare critiques of the consultation process, as reported by the project staff;
- to encourage community planning for improved transport conditions for a small country town;
- to undertake a study of a major traffic corridor; and
- to help project teams responsible for transport to change their management style.

Waste management
Paul Cosgrave
One of the most contentious environmental issues of 1990–91 was the process of determining a site for a high temperature incinerator in rural New South Wales. The debate became progressively more bitter as communities and the Commission investigating various options became cast into opposing roles. Questions that were asked at the time of the enquiry included:

- What factors should determine the siting of the incinerator?
- What role should the community have in the investigation process?
- What would be the effects of the incinerator on nearby residents' health and the local environment?

This project was aimed at bringing together the goals of the communities affected by the decision and the goals of the Commission in finding a suitable site for the incinerator.

Remedial conflict management
Clark Ballard
While some environmental issues do appear literally overnight – for instance, chemical blazes and oil spills – most have a much more protracted development, dragging on for years or even decades. Individuals and communities can become entrenched in positions which become increasingly insulated as time passes. Such positions can become immune to even the most severe community pressures. Major remedial action is needed to bring conflicting interests together to discuss the issue. This project was aimed at addressing such an issue in a Victorian community.

Conflict management in environmental law
Margaret Sidis
The legal framework that operates in Australia is highly oppositional in nature, with the prosecution and the defence etc. As environmental conflicts become resolved through the law courts, there is every likelihood that Environmental Law could acquire this adversarial process, instead of adopting more positive conflict management techniques. This project was aimed at introducing different strategies – such as negotiation and win-win outcomes – to the legal profession.

The development of a conflict management sourcebook
Valerie Brown
As part of an ongoing project at the Centre for Resource and Environmental Studies, a conflict management sourcebook was be developed. The Sourcebook will be used in conjunction with further courses to be offered by the centre, and also as the basis for an undergraduate course in Environmental Conflict Management. Specific management issues that needed to addressed by those developing the text were:

- to achieve equal contribution from, and responsibilities for, all members of the project management;
- to gain acceptance amongst stakeholders in the community for the Community Management Group as a legitimate agency.

Navigational channel in a coastal creek
John Downey
This project in a small coastal town has been a source of long-standing community conflict. A community diagnosis and needs analysis had been carried out, but none of the parties – Shire Engineer, real estate developer, and community interests – knew what to do next. The strategies in Sessions 3–7 were used to negotiate, implement and evaluate a programme which met the needs of all parties.

Appendix 1
Course Coordinator's Guide

BACKGROUND

Risks and Opportunities provides materials for those learning to manage environmental conflict and change. Participants in the programme could be students, colleagues, executive management or national agencies learning to work together.

In every case, a course will require the identification of a coordinator – someone who takes responsibility for the organisation of the programme and the learning environment of the participants. The coordinator of the course will need to:

- select from the materials to meet the needs of their particular issues and clientele;
- adapt the materials to the personal skills and styles of course participants;
- have previous experience in working with groups; and
- have an understanding of self-directed learning and learning by doing.

Materials

The three sets of material in the manual are intended to be used together, as theory, learning and practice. In fact, practising what is preached.

Part 1 **Theory**: A background paper provides an overview of the management issues in environmental conflict.

Part 2 **Practice**: Four case studies of different patterns of environmental conflict in three countries provide source materials for the learning modules. They can be applied at the depth required by any particular course.

Each case study has a matching simulation exercise in the learning modules designed to enlarge the experience of the course participants.

Part 3 **Learning**: Eight learning modules allow for the exploration of the conflict management issues using the experience of the course participants themselves.

Key Issues

Key issues in managing environmental conflict which the coordinator will need to be prepared to develop throughout the course are :

- **Leadership** – coming to terms with the various roles of expert/manager/decision-maker/facilitator;
- **Integration** – the need to synthesise expert advice, community priorities and economic projections into a coherent form;
- **Time** – the understanding that planning, team-building, consultation, personal learning and organisational change all work at different tempos, and that each of them take time to come to fruition.

Timetables

The course as presented in the manual is run over eight sessions, with timing appropriate for a five-day course, with each session allotted half a day (Timetable A). The materials can be adapted to a weekend in-service programme (Timetable B) or a semester academic unit (Timetable C). They have been trialled as each of these, as a personal learning programme and for management training. The practical application of the materials in the workplace setting is described in Section 3.4.

	MONDAY Dec 2	TUESDAY Dec 3	WEDNESDAY Dec 4	THURSDAY Dec 5	FRIDAY Dec 6
9:00	*Pre-reading materials despatched week before Workshop*	1. EXPLORING - FORECASTING Review of program	3. GOALSETTING - MUTUAL ADVANTAGE Plenary: Getting to Yes: Principles of Negotiation	5. EVALUATION - ILLUMINATION Plenary: Who is judging? Who is judged?	7. LEARNING - ORGANISATIONAL CHANGE Plenary: Supports and Barriers
9:30					
10:00					*"What is the game?"*
10:30					
11:00		BREAK	BREAK	BREAK	BREAK
11:30		Using Vision	The Harvard Process	Performance criteria	Strategies for change - Closing session
12:00		*"All the issues on the table"*	*"All the players around the table"*	*"All the players know the game"*	CLOSE
12:30	Registration				
1:00	LUNCH	LUNCH	LUNCH	LUNCH	LUNCH
1:30		2. ANALYSIS - COORDINATION	4. ACTION - COOPERATION	6. LEARNING - INDIVIDUAL SKILLS	
2:00	INTRODUCTION Welcome *Professor Henry Nix*	Players and Powerbases	Management Styles	Leadership styles	*Individual Project - consultations with Workshop Staff*
2:30	Risks and Opportunities *Val Brown D. Ingle Smith John Handmer*	*"All the players in the game"*	*"All the interests in the game"*	*"Who are the players?"*	
3:00					
3:30		Discussion	Discussion	Discussion	
4:00	BREAK	BREAK	BREAK	BREAK	
4:30	Expectations and experiences *Linden Orr*	Agreeing on the agenda	Agreeing on objective criteria	Applying the skills	
7:00		INFORMAL DINNER		WORKSHOP DINNER & HYPOTHETICAL	

Figure A1.1 Risks and opportunities – a week-long workshop timetable

Friday
6:00–9:00   **Introduction**
                                   Conflict management studies
                                   Developers of Id
                                   Supper
Saturday
9:00–11:00   **Session 1**
                                   One forecasting exercise
11:30–12:30   **Session 2**
                                   Stakeholder analysis
1:30–3:00   **Session 3a**
                                   Review of Harvard Negotiating Program
3:30–5:00   **Session 3b**
                                   Identify priority goals
Evening   **Course dinner**
                                   Hypothetical
Sunday
9:00–11:00   **Session 4**
                                   Identify management styles
                                   Triads
11:30–12:30   **Session 5**
                                   Evaluate stages and strategies
1:30–3:30   **Session 7**
                                   Organisational change

Figure A1.2 A weekend workshop timetable

Week	Seminar (2 hours)	Practical (3 hours)
1	Issues paper: Why, What, How and Who	Analysis of press reports of current environmental issues
2	Water resource management: Case study of salmon	Guest speakers from a range of environmental management roles
3	Conflict resolution in natural resource management	Conflict resolution styles – personal and organisational
4	Case study of Nyah Shire	Hypothetical
5	Case study of Bangkok	Student short papers on market mechanisms
6	Case study of Calico Creek	Exploration and Forecasting (Learning Module 1)
7	Community led land and water use planning	Analysis and Coordinating (Learning Module 2)

(*Semester break*)

8	Harvard negotiation Process 1	Negotiation Exercises
9	Harvard negotiating Process 2	Goalsetting and Coordinating (Learning Module 3)
10	Negotiating water (Handmer 1991 et al)	Analysis of management styles in text and in class (Learning Module 4)
11	Environmental Impact Assessment	Student critiques for criteria for environmental monitoring
	Review of state of the art	Valdez Principles (Learning Module 5)
12	Components of change – individual and organisational learning	Diagnosis of individual communication and learning styles (Learning Module 6)
13	Social change models and styles of social change	Factors enhancing and and inhibiting organisational change (Learning Module 7)
14	Reports of learning groups	Evaluation of Course

Figure A1.3   A semester unit (14 weeks)

SETTING THE AGENDA (CHAPTER 10)

In the introductory session, the course coordinator gives participants a sense of where they are, where they are going, and where they will be at the end of the programme. The issues paper helps here, since it identifies each of the environment and management issues to be raised in the course. Ideally Chapters 1–9 should be read by course participants before the introductory session. In the real world, this hardly ever happens and it is necessary for the course coordinator to provide a short overview of:

- the issues paper *Why, What, How and Who* (Chapters 1-5);
- the principles of negotiation; and
- the four case studies (Chapters 6–9).

Thirty-minute discussion time should ensure that all participants have a chance to discuss their interests and issues.

The course coordinator will ask participants for their own ideas and experience on conflict management issues. Allow time for each person to introduce themselves, their background, and their learning objectives. Records should be kept of the group's objectives for review at the close of the course.

The two activities in Chapter 10 introduce the participants to the ideas:

- that there are a range of conflict management styles which we all use, and which are all valid in some circumstances; and
- that any form of environmental management has inherent conflict, which can cause high levels of anger and confrontation unless identified and clarified.

It is essential to leave a full half-hour for discussion on each exercise. During that half hour, as coordinator, you should ensure that:

- every participant contributes;
- every contribution is valued;
- feelings of uncertainty and anger have the opportunity to be voiced;
- discussion responds to questions of clarification, not criticism of other's ideas;
- there is no 'good' or 'bad' response, only contributions to discussion;
- you sum up each exercise and review the outcome with positive style, confirming the points made above;
- you introduce the rules for debriefing participants after each simulation exercise.

Throughout this guide, the coordinator is expected to have the background knowledge and skills indicated up to this point. They are not assumed to be expert in any particular area of environmental or conflict management, but to be capable of skilled coordination and facilitation of self-directed adult learning.

Debriefing

The course coordinator bears the responsibility for the positive learning of all of the participants in the course. By using simulation exercises, the possibility of practical learning is increased – but so is the risk of negative learning. Negative learning follows where the learner has had a different experience, has not been given opportunity to work the issues through for themselves, or has been given a model they do not want to follow.

The discussion during and immediately after each exercise is vital in avoiding these types of learning blocks, and ensuring that the learning is positive for all participants.

The course coordinator will need to debrief the participants by ensuring that every participant:

- has the opportunity to describe how they felt doing the exercise, as well as what they learnt;
- comments constructively on their interaction with the others during the exercise;

- discusses any apparent differences or debates;
- closes the exercise brief clearly separating themselves from their experiences during the exercise by
  - removing name labels allotted during the exercise,
  - discussing the implications of the exercise 'for the real world', and
  - identifying the difference between their experience in their occupations and in the exercise.

If these opportunities are not provided, learning is not only blocked, but may reject the very skills in conflict management which the course attempts to develop. When they are provided, learning follows the 'adult learning cycle' outlined in Chapter 1.

CASE STUDY SCENARIOS: CHAPTERS 10–17

The four case study scenarios briefing sheets, given below, provide a background for the exercises in Chapters 10–17. Three lengthy detailed simulation exercises are provided which can really challenge participants' conflict management skills in Chapters 11, 12 and 13.

Case Study A: A Heroic Effort: The Salmon Fishing Industry of British Columbia (Synopsis of Chapter 6)

Summary

The Fraser Basin has an area of 234,000 km^2 and supports one of the world's largest salmon fisheries. The numbers of migrating adult fish have declined by some 80 per cent since the turn of the century. Major conflicts have existed for decades among a range of stakeholders. The recognition of the rights of indigenous people is a significant, but recent, addition to the conflict.

Characteristics

1 **Policies**: the salmon fisheries are managed by a mix of federal and provincial agencies; international treaties, management contains elements of economic and scientific approaches; federal legislation for native peoples.
2 **State of the environment**: river and estuary pollution from a variety of sources; including urban runoff, industrial waste (especially from production of wood pulp), and forestry practices; effects of solutes

and suspended sediment; changes to flow regime from transport construction and potential from construction of dams.

3   **Community interests:** federal and provincial governments; fishing groups (these include commercial, recreational and native peoples); metropolitan, rural and urban communities; pulp wood industries; forestry; fish biologists; conservation groups.

4   **Individual skills:** international treaties; federal and provincial judicial and bureaucratic mechanisms; coalitions of native peoples; co-operative management; market mechanisms; scientifically-based resource management.

5   **Agencies and services:** Pacific Salmon Treaty organisation (with the USA); Federal Department of fisheries and Oceans; various Provincial Ministries; metropolitan and local governments; Fraser River Estuary Management programme, Indian Fisheries Commission, commercial fishing groups such as the Pacific Salmon Defense Alliance; various union organisations, including United Fishermen and Allied Workers Union and the Native Brotherhood of British Columbia; environmental groups such as the British Columbia Wildlife Federation.

Case Study B: Nyah Shire: Repairing the Damage (Synopsis of Chapter 7)

Summary

Nyah Shire covers some 1500 hectares on the Victorian banks of the river Murray, with a stable, ageing population of some 1800. The district is badly salt-affected from long irrigation on small sized holdings. Only 14 per cent of the district's farmers are full time and the water is heavily subsidised.

Characteristics:

1   **Policies:** water is managed as a 'public good', an uncosted community resource; method is bureaucratic-consultative management, with community, state and federal interests in partnership. Pollution levels are, however, increasing to the point of non-viability of the land.

2   **State of environment:** water is abundant, untreated controlled, polluted and subsidised; water use is half national irrigation supply; salt-polluted with degradation increasing; agricultural grazing and household uses; 1500 inhabitants.

3   **Community interests:** upstream and downstream irrigators; farmers; graziers; business community; householders; environmental

scientists; conservation groups; full range of political party membership.

4 **Individual skills**: formal consultation mechanisms, lobbying, organisational management; agricultural management; conflict resolution methods.

5 **Agencies and services**: Rural Water Commission; Murray-Darling Basin Commission; Community Advisory Committees; Local Action Advisory Groups; Victorian and New South Wales Departments of Agriculture, Health, and Conservation and Environment.

Case Study C: Trouble with Traffic – Effects of Rapid Urbanisation in Bangkok (Synopsis of Chapter 8)

Summary

The severity of traffic congestion in Bangkok is of world renown. Rapid industrialisation has escalated Gross National Product, and left severe by-products including a traffic impasse where the journey to and from work takes eight hours a day. The average speed on main roads is below 15 kph and less than 5 kph in the central business district. The pollution from all forms of vehicles exceeds WHO standards and adversely effects the physical health and well-being of the 6 million residents. For many years water transport on the Chao Phraya River and the klongs decreased but is now increasing in an attempt to improve traffic flows.

Characteristics

1 **Policies**: Bangkok Metropolitan Administration, separate industry, health, finance, education departments; traffic police, international aid; global energy policies.

2 **State of the environment:** air, noise and water pollution are chronic, all exceed WHO standard at times but difficult to document; atmospheric particulate matter, lead and carbon monoxide are of special concern; aquatic life in the rivers and canals is non-existent; 21 per cent of vehicles and 80 per cent of boats exceed Thai noise standards.

3 **Community interests:** planners, traffic police, commuters (bus, car and motor cycle), non-government agencies; Buddhist groups; environmental scientists, universities, industrialists, social welfare agencies, health officials, families, expanding poverty, displaced rural workers.

339

4 **Individual skills**: Traffic engineering, city planning, political lobbying, community organisation, traditional stoicism and meditation, survival skills in long-term (four-hour) daily commuting, social change skills.

5 **Agencies**: Bangkok Metropolitan Administration, national government departments (eg Department of Industrial Works, Office and Town and Country Planning) NGO's (including Buddhist religious/political groups), industrial organisations.

Case Study D: Calico Creek: Managing the Commons (Synopsis of Chapter 9)

Summary

Calico Creek is a small creek supporting a dozen farmers in Central Queensland. Most of the time there is more than enough water; and the farming is small scale and traditional. Every few years there are drought conditions, and the amount and allocation of the water become critical. There is also a recent movement to diversify crops and to reach larger markets. A local Catchment Advisory Group of water users manges the quotas in good times and bad.

Characteristics

1 **Policies**: water is managed as a 'commons' in which all users nominally have equal rights; a community committee agrees on protection against extreme events, in this case, drought and flood. There are pressures for increased productivity and economic development.

2 **State of environment**: water is abundant, untreated, uncontrolled, unpolluted and uncosted; subject to extremes of supply; used for small crops and dairying in farms ranging from 5–300 acres; minor household use; 60 inhabitants.

3 **Community interests**: users (in the widest sense) are farmers, graziers, irrigators, local households, consumers of agricultural products, local self-sufficiency; regional developers; National Party membership.

4 **Individual skills**: neighbourly negotiation; committee management; agricultural management; rural resourcefulness.

5 **Agencies and services**: Water Resources Commission; State Departments of Primary Industry, Health, and Conservation and Heritage; Australian Environmental Council; Heritage Commission.

The three simulations in this sourcebook have been carefully graded.

Irrigation Inc: represents the most structured of the three – each character's positions, issues and options are carefully delineated. There are only a finite set of solutions available through them, and a limited number of issues under consideration. Conflict will tend to turn up in the form of interpersonal conflict and conflicting views that have been written into each part.

The hypothetical – Nyah Shire: represents an intermediate phase. Each participant has a thumb-nail sketch of their character, within which they are free to extemporise. However, the participants will have to decide what they see as the issues and their priorities. There is a good deal of conflict possible within this simulation. The simulation requires at least sixteen people to play – preferably several more. People do not work effectively in one-to-one within such large groups. The appearance of factions and interest groups is inevitable.

There is a very large amount of information to deal with – probably more than the average person can easily cope with. In the heat of the debate, details are going to become blurred. There are also going to be different interpretations, and misunderstandings – exactly as happens in the real world.

The role of the moderator needs special mention. The moderator is keep the story moving, and should not interfere in the story itself. Only if the entire story is about to break down because of conflict should the moderator intervene. The participants are the ones who must management the conflict – this is why they are doing the course.

This simulation was originally written as an after dinner 'entertainment' where participants could participate in a convivial, and light-hearted atmosphere. This approach will short circuit much conflict, and place the exercise in a non-confrontational setting

The open forum on Bangkok has the loosest structure. The participants themselves have to decide the people that should be involved in the discussion, and what the issues to be discussed are. How the discussion is run is up to the participants: they may elect a chair, they may elect a panel, they may caucus amongst themselves, they may have a free-for-all.

In preparing the materials, it has been assumed that whoever is responsible for the course will:

- select critically from the materials to meet the needs of their particular issues and clients;
- trial the materials with colleagues and/or clients;
- have, or have access to someone who has access to, previous experience in working with groups; and
- have a commitment to self-directed learning and learning by doing.

Note: participants should not be asked to reveal their scores, although they may wish to do so themselves.

Calculating each person's score will take about 20 minutes. There should be at least 20–30 minutes of discussion about the implications of the personal scores. The following points should be made:

- all types of conflict management are appropriate at some point;
- everyone uses all five responses at some time;
- the ideal is to choose the appropriate response rather than fall into a familiar pattern

Participants may like to discuss their own experiences with each conflict resolution style and types of environmental conflict for which each of the five types is most suitable.

EXPLORING AND FORECASTING (CHAPTER 11)

General

In Session One the challenge to the course coordinator is to provide a constructive creative environment for three consecutive exercises. The session offers three styles of future searches, in decreasing emphasis from open-ended to task-focused. The first exercise focuses on the changing social context of a selected issue; the second on the future resolution of the issue, and the third on the current key components for management. The task is to maximise the participants' contributions; and for each exercise to build on these in turn.

The course coordinator should consider the following options.

1 If the client group is highly structured, and/or the majority of members have convergent thinking styles, a more structured exercise than 'Future World' could be used. The script for the Issues Analysis exercise can also provide the background material for the 'Future World' exercise.

2 If it is a short course (two or three days) then only one of the search exercises should be used. Which exercise is chosen is a matter of matching the clients' objectives. A task-centred group may need the first exercise in order to expand their options. A group that has been considering policy options for some time may need to focus on components of implementation in the third exercise.

3   If the programme is an academic unit, the three exercises can be presented weekly, combined with material from the Course Reading.

Materials Available

- Issues analysis
- Future world
- Visioning the future
- Futures Wheel

ANALYSING AND COORDINATING (CHAPTER 12)

General

In Session Two the work of the course coordinator is to ensure that each participant increases their level of skill in selecting and coordinating a management team. This applies whatever their actual role in environmental management. It is assumed that every player in an environmental issues has a legitimate stake in the management of that issue. The advice in Chapter 11 of Session Two applies to the course coordinator as well as to participants. Resources available for the course coordinator for Session Two are:

- *Getting to Yes*;
- Course guide on meeting procedures; and
- Course paper on coordinating the stakeholders.

Skills needed by the course coordinator are experience in managing role play and the essential debriefing which follows.

Note: It is essential to provide name labels for each role adopted by a participant. Players should stay in the role throughout the simulation exercise (they can be asked 'are you speaking to me in or out of role?' if there is any uncertainty). At the end of the exercise, the coordinator should ensure that each person takes off their label and says something like 'I am no longer the mayor: I am Jane/Fred'. After removing their label, each person should be asked if they have also got rid of the feelings which went with the role. If not, they should express these. The simulation exercise should conclude with participants expressing appreciation of each other's contribution in role.

Materials Available

- Stakeholder analysis
- Stakeholder resource analysis
- Setting up a committee
- Preparing a policy, strategy or action

SETTING GOALS AND NEGOTIATING (CHAPTER 13)

In Session Three, the course coordinator will need to decide whether participants will continue in the roles selected in Session Two, or choose new roles. The decision will rest on the participant's own choice, the degree of learning in Session Two, and indications as to whether the roles are still capable of development.

In Session Three, the participants can use any case study of environmental management to supply the information for a longer simulation of decision-making in environmental management. There are scripts provided, or participants could prepare their own. In order of complexity they are:

- **Script 1**: Hypothetical case based on Nyah Shire;
- **Script 2**: a Stakeholder simulation based on Trickle Creek;
- **Script 3**: Public consultation in Smoke City from which participants develop their own roles.

Note: It is essential to provide debriefing after this session. See Session Two for details.

TAKING ACTION IN A COOPERATIVE TEAM (CHAPTER 14)

General

One of the more fruitful approaches to negotiation in recent years has been developed by a team from the Harvard School of Business Management. Although criticised as being too much of a recipe, and as too market-oriented, the Harvard Negotiating Process contains basic principles which can be adapted for positive resolution of almost any issue. The four steps of the Harvard Negotiating Process are:

1   separate the people from the problem;
2   focus in interests, not positions;

3 invent options for mutual gain; and
4 arrive at objective criteria.

Course participants have by now experienced some of these skills in Sessions 1 and 2. They have also reviewed the case studies of water resource management in Session Three. At this stage, the coordinator could revise the four steps listed above noting that each step is as necessary in everyday interactions between team members and between team members and clients, as they are in committee negotiations between interest groups.

Effective cooperative teams require that each member shares both the leadership risks and the responsibilities. An effective team will contain a range of styles and skills which, working together, can contribute much more than the sum of all the individual's contributions. The course coordinator may like to present the attached notes on leadership as a short seminar or lecture.

This session requires the course coordinator to:

- administer a diagnostic test identifying leadership styles and discuss the implications of the results (30 minutes);
- supervise reflective listening triads (30–60 minutes); and
- organise a goldfish bowl demonstration of negotiating as a delegate (20–40 minutes) (optional).

Materials Available

- Problem-solving style
- Cooperative team leadership
- Representing the team
- Leadership style

EVALUATION AND ILLUMINATION (CHAPTER 15)

The group brainstorms the dimensions of evaluation. Illuminative evaluation:

- collects quantitative and qualitative evidence;
- makes progressive and summative (final) assessments;
- works with external and internal observers of the project;
- monitors intended and unintended outcomes;
- describes task outcome and management processes;
- is both interpretive and objective; and
- distinguishes between certainties and uncertainties in prescribing the results.

Local methods of state of the environment monitoring, rapid rural appraisal and integrated environmental impact assessment should also be revised.

The course coordinator can assist group members to discuss their previous experience of environmental impact assessment, and their expectations of future assessment. The group could brainstorm the many stakeholders with an interest in the evaluation. The essential stakeholders include all decision-makers in the area of the selected environmental issue. Discuss through what channels of communication does each decision-maker receive the evaluation results? Will the evaluation answer the questions each stakeholder wants answered?

This is the final meeting of this group as a group. Members should evaluate their own work individually and as a group. The group is asked to prepare a description of their case study solutions and their problem-solving process to present to other participants at a plenary. Before finishing, it is important that group members acknowledge:

- contributions of each member;
- insights into substantive issues;
- insights into conflict management;
- insights into group functioning.

INDIVIDUAL LEARNING AND PERSONAL CHANGE (CHAPTER 16)

General

This session is on a personal level. Participants will need a safe and constructive atmosphere if they are to discuss the different score sheets on individual styles of conflict management, group leadership and group roles. Forming new sets of triads are one good way to organise this. An introduction from someone who works in personal development is a useful resource for this session. If someone trained in this area can take charge of the session, even better.

The session allows participants to re-examine personal goals and priorities as illuminated during the course. The individual objectives tabled by the participants in the introductory session can be reviewed here. Evaluation of past and future learning could end with each participant completing an adult learning contract for themselves. (see the form in Session 2.9)

As Course Coordinator, you might like to choose to run exercises in communication, for instance a public communication, the channels for transmission and reception (NLP exercise); and group communication. If it seems appropriate, further diagnosis of thinking and learning styles could add to understanding of how a common solution can be found within diverse groups and interests.

Myer-Briggs is one form of analysis, another is Kolb's analysis of problem-solving styles. The notes in Session Six on specialist, generalist and holist can be discussed in relation to participant's roles in their workplaces and environmental management teams.

Materials Available

- Conflict management style
- Negotiating style
- Group leadership style
- Group roles

ORGANISATIONAL LEARNING AND SOCIAL CHANGE (CHAPTER 17)

General

The most difficult stage of problem-solving is not as many expect, achieving new, effective and negotiated solutions. This has been the focus of the previous six learning modules, and is difficult enough. It is even harder to make those solutions effective within an existing community or organisation. Incorporating the solution into standard practice in existing organisations and communities is a very real challenge.

Managing the Future (45 minutes)

Force Field Analysis is a technique in which participants can identify factors which encourage and discourage the acceptance of change within their own organisations. This needs to be managed very constructively, with participants working together on common problems and common solutions. The ideal is for participants to continue to provide a support group for one another after the programme is finished.

Section Three of the Sourcebook contains ten case studies of the successful application of the environmental conflict management strategies in a wide range of situations, from engineering projects to peace studies.

Sharing Learning (45 minutes)

Group work as a plenary: the three working groups formed during the previous learning sessions share their conclusions on:

- general principles of environmental management for the future;
- general principles of conflict management for environmental decision-making; and
- the practical implications for their workplaces.

Finally there will be a course evaluation – and refreshments. Participants will be asked to fill in a comprehensive course evaluation.

Coordinator's Tasks

1 Force-field analysis involves drawing horizontal line across a whiteboard or a large sheet of paper as in the diagram.

Participants are asked to identify a specific project on environmental conflict management which will be of value to them to test the learning from the workshop when they get back to work. They could do this individually or in small groups whose members assist one another. When each participant has identified their own projects, participants individually, in small groups or as the whole group, brainstorm the inhibiting and enhancing factors which will affect the implementation of the project back at their own work. The group then works as a team, suggesting ways in which enhancing factors can be strengthened and inhibiting factors reduced.

It is up to the coordinator to decide the value of the participant's working as individuals or as groups. Once decided, the pattern should be maintained until the exercise, which should always end with the full group sharing their expectations in returning to work.

2 The course coordinator would summarise by asking each group to appoint a chair and reporter. The reporter would summarise the three main points under each of the suggested topics on an overhead transparency, or on a large piece of paper. Then the course coordinator could bring the groups together in a plenary to present their conclusions to each other. After a 15–30 minute discussion, the learning session could proceed.

Discussion

This exercise requires 30–60 minutes discussion time to allow participants to share their learning and to form cooperative partnerships to assist each other in the future.

Finally, the programme should close with time to 'have a party' say goodbye to each other, and for support groups to arrange to meet again.

Index